AF777520

BROADAX ARTISTS

BROADAX ARTISTS

George B. Linn

Janevar Publishing Co./Muncie, Indiana

Copyright © 1984 by George B. Linn
All Rights Reserved

ISBN 0-937174-02-5
Library of Congress Catalog Card Number 83-83044

Published by Janevar Publishing Co., R. R. 11, Box 129,
Muncie, Indiana 47302

Printed in the United States of America

TO DAD AND MOTHER

Contents

Preface

I remember my first glimpse of a tie camp on a late afternoon in November 1919. After traveling by sled over vast snow-covered areas, my family and I entered the Bridger National Forest. As we descended a steep hill I saw a small valley below where a dozen or so log structures nestled among the pine trees, smoke poured from their stove pipes, and huge snowdrifts were everywhere. This site on Cottonwood Creek was the headquarters of the tie camp operated by the Standard Timber Company of Nebraska about thirty miles north of the little town of Big Piney, Wyoming. Founded in 1912 by Omaha residents Dan Wilt, N. P. Updike, and the Woods brothers, the company supplied ties to the Union Pacific Railroad.

Summer visitors loll today in the shade of the pine trees in a picnic area beside a road which runs through the valley. They are probably oblivious to the fact that this space was once occupied by a huge commissary in which was stored a supply of groceries, work clothes, and tools. The commissary was one of a row of log buildings which included a blacksmith shop, several bunkhouses, and a barn. No sign of them remains today. Ranging in age from ten to fifty years, a stand of trees has replaced the home of Bror Edin, the camp manager, and his family. Grass and shrubs grow where once stood a small garage and a cookhouse which fed as many as twenty-five hungry timber workers. Cottonwood Creek which is shallow and narrow in summer flows beyond. A summer visitor might find it hard to believe that during the spring thaw the water in this creek had enough force and depth to float thousands of railroad ties down the stream to the railroad.

Graded roads now follow both the north and south forks of Cottonwood Creek. They have replaced the primitive trails which existed sixty years ago. A United States Forest Service official patrolling the region will courteously greet a traveler, give him a map of the area, and point out Lander's Peak, Soda Lake, and other landmarks. He is unaware that the open space in the forest just below Soda Lake was the site of the Burns camp during the 1920s or that Alvin Hardin's camp existed deeper in the forest on the opposite side of the creek.

In addition to the headquarters, commonly known as the commissary camp, other camps were scattered throughout the forest from two to five miles apart. Each had a cookhouse, a barn, and a few bunkhouses. Some of the tie hacks lived alone in their own cabins. A few had families. The majority of the tie hacks and teamsters lived in bunkhouses and got their meals in cookhouses. We were to make our home in the Moline camp about three miles from the commissary.

Now that it is behind me, I regard my boyhood in tie camps as a richly rewarding adventure. I have often longed for the serenity of a Wyoming forest miles from the beaten path. There is another side of the story, though. I remember the beauty of the vast expanse of pine trees surrounded by snow-capped mountains; to the tie hack, the forest was a place to earn a living the hard way. At times I have longed for the winter blizzards, the sleigh rides, the tobogganing, and the skiing; to the tie hack, snow and storms added to his burden and the only good thing about sleds and skis was that they beat walking through deep snow. A quiet evening beside a roaring fire in a Franklin heater and the smell of wood smoke gave me a feeling of ecstasy; to the tie hack, his cabin or bunkhouse was a place to grab a night's sleep before getting up at six o'clock in the morning for another ten hours of grueling work. Camping out and sleeping in a tent was great sport for me; to a man on a tie drive who waded all day through cold water while tugging at ties, eating in the open regardless of weather, and sleeping in a "rag house" only made life rougher.

The tie hacks took their work and way of life in stride and usually remained amazingly cheerful and retained a sense of humor. Without the latter, the isolation, backbreaking work, primitive living conditions, and hard knocks might have driven them berserk. In order to make ends meet, they adhered to a tiring schedule; but they found time for a little fun and frolic. Something which caused a good laugh frequently occurred; if it did not, they created such a situation.

The tie camp deserves its place in history along with ranch life, the cattle roundup, the mining towns, and other phases of life in the western United States which are rarely portrayed accurately in either fiction or nonfiction. I am glad to have an opportunity to put on record my recollections of those hardy and courageous broadax artists and their families who endured a way of life which now belongs to the past.

Some information in Chapter 10 appeared originally under the title "The Drivers of the '20s," **In Wyoming,** VI, No. 1 (April-May, 1973), pp. 20-1, 31.

One's memory of experiences during his youth is, of course, not infallible. Wherever possible I have checked events and dates with records and have conferred with others. My mother, Laura Osborn Linn, was a valuable source of information. I am indebted to Nils Edin for information regarding the organization and early history of the Standard Timber Company and for permission to reproduce his photographs. I am also grateful to Linnea Moline Damon and Andrew Moline for providing useful information pertaining to the operations of the Standard Timber Company in the 1920s.

CHAPTER ONE

Early Years in Wyoming

My grandfather, John Bartow Linn, was born on a plantation near Cartersville, Georgia, in 1861 shortly before the outbreak of the Civil War. His father later returned from a Yankee prison camp to find that his slaves had run away, his livestock was gone except for one old horse, and that no crop had been planted. Not long after that he loaded his family and all of his remaining possessions into a wagon and moved to the wide open spaces of Arkansas where there is now the thriving town of Mountain Home. Granddad grew up and married in Arkansas. It was there that my father was born.

While still a young man, Granddad Linn was told that he had consumption and needed plenty of fresh air and exercise. No one knows, of course, what diagnosis would be made today of his condition, but he lived to be eighty-five years old. He went to Colorado and plowed ditches, waded through mud and water, and slept on the damp ground. The life seemed to agree with him, and he had his family join him. I have been told that he had gained so much weight that Grandmother Linn scarcely recognized him. My father was a young child when his family settled in Rocky Ford, Colorado. He grew up there, and at an early age he began working in sugar factories in Rocky Ford and Scottsbluff, Nebraska.

My grandparents moved to Wyoming prior to 1910. Grandmother Linn died in 1912. Granddad homesteaded a tract of land at the foot of Fall River Basin, now known as Hoback Canyon. Later he bought a small ranch near Daniel, Wyoming, and over the years acquired a herd of cattle.

Granddad was fond of children. He had much affection for his grandchildren, and eventually he had fourteen of them. Probably all of our parents felt that he spoiled us. It was always a thrill to take a ride with him in his wagon, and he never missed an opportunity to have one or more of us with him. Until I was six years old, he had a home in Marbleton which is a little town north of Big Piney. In the late summer he would go to his homestead cabin in the basin and remain there until fall. It was always a thrill to have old bewhiskered Granddad back again. While on his ranch, he ceased shaving and would let his whiskers grow until spring. Then he would clip and shave his beard, retaining only his huge moustache. I never knew what he did on the homestead other than hunt big game. He never failed to bring several quarters of elk meat when he returned.

He was not a big man. About five feet nine inches tall, he was stocky but could not be called fat until he reached the age of sixty-five. His appearance was not that of an athlete, but at picnics and outings during his younger years he frequently demonstrated considerable skill in sports, especially in broad jumping.

Although Granddad lived in the north most of his life, he remained at heart an unreconstructed rebel. While I do not believe I ever wholly accepted the logic of some of his arguments regarding the Civil War, I am afraid that a few of his prejudices rubbed off onto me which I retained until I was a junior at the university. Throughout his life Granddad's pronunciation remained definitely southern, and his speech was sprinkled with such expressions as "som'ers" for somewhere, "Ah cain't he'p it," "thataway," and "thisaway." If he especially liked something, he would say "Ah think a heap of it."

My mother, Laura Adelaide Osborn, was born in Clifton, Illinois. Her father was born in the United States to English immigrants, and her mother had immigrated with her parents from England to the United States. When Mother was a small child, her family moved to the vicinity of Mound City, Kansas, where her father acquired a farm. Grandmother Osborn died of pneumonia when Mother was sixteen years old. Shortly thereafter, Mother went to live in Rocky Ford, Colorado, with her older sister and started working as a telephone operator. It was there that my parents were married in 1911.

In 1912 Dad and Mother decided to move to Wyoming. Dad's parents were then living in Boulder. My parents traveled by train to Rock Springs where they arrived on a Saturday evening and spent the night. They encountered mine workers and others who were enjoying a payday by thronging noisily in and out of the many saloons. My parents had a room at the Valley House, probably the only hotel in Rock Springs. Mother has told me that she lay down on her bed and wept, and Dad told her to pull herself together. Arriving in Boulder, they found that the town consisted of a general store, a few residences, and little else. To get there, they had ridden on a light truck. The road was wet and muddy, and there were steep hills to climb. From time to time it was necessary for passengers to get out and push the truck in order to help it over a hump or through a mudhole. Attired in a suit of clothes appropriate for the well-dressed young man of the day, complete with Oxfords and a derby hat, Dad soon learned that he was not properly clothed for Wyoming fall weather.

Dad worked as a ranchhand for a couple of years. I was born on Henry Clevelin "Doc" Dunham's ranch near Big Piney in 1913. The following year we lived on the Bird place located between the tiny railroad town of Opal and Big Piney. I do not remember life there but was told later that the ranch had one of the more up-to-date houses in the area. Mr. Bird had contracted to carry the mail from the railroad to the Piney country. Dad's job was to take care of the ranch and feed and bed the horses of travelers to and from the railroad who used the ranch as a stopover.

Anticipating the arrival of my brother Bill, we moved to Marbleton in the summer of 1915. Instead of relying on a midwife's services as was customary in that time and area, my parents arranged to have a doctor in attendance. Soon after Bill's birth we moved to Big Piney, and it was at that time that Dad began freighting. In 1916 Dad and Granddad leased the Starmont place near Big Piney. Dad continued to spend part of his time freighting while we were on the ranch. It soon appeared that he could make more money by freighting than by operating a leased ranch. We moved to Marbleton where Herschel was born in 1917. Dad continued to freight. Herschel remained the youngest member of the family until James was born in Evanston in 1927.

Freighting provided a comfortable living, but it was a grueling occupation. The road consisted of nothing but wagon tracks, and during much of the year travel with heavy trucks was impossible. A light truck carried the mail between Opal and Names' Hill, and thence it was relayed to another light truck between Names' Hill and Big Piney. Heavy freight had to be transported by horsedrawn wagons.

Dad hooked two wagons together and hitched six head of horses to the front wagon. Seated on his bedroll near the front of the lead wagon, he made the long trek to the road, yelling at horses by name, "Yah, Rock! . . . Giddap, Maude! . . . Carrots! . . . Giddap there!" It did no good to yell at Shorty who was very hard of hearing. If Shorty lagged, Dad cracked his whip near the former's ear. This woke him up and kept him moving. Occasionally Dad threw a rock at a horse from a supply he kept in the wagon for that purpose. He drove until nightfall.

After unhitching and unharnessing the horses and putting nosebags of oats on them, he pitched his tent and made a fire. Taking his box of grub from the lead wagon, he cooked and ate his supper. He then turned the horses loose with a bell suspended from the neck of one of them, and he turned in for the night. Early the next morning, he got his breakfast, hitched up the horses, and drove on.

Dad's route passed a number of ranches, including the Bird place, McGinnis's, Haddenham's, and others. For certain long stretches, though, there was nothing to see but miles of barren land covered with alkali. It was essential to gauge each day's drive so that at the end of the day he could stop where there was water for the horses and a supply of green grass during the summer. Dad drove continually each day, except for a short stop to prepare and eat his noon meal. After reaching Opal he loaded both wagons with freight for the stores and ranches in the Piney country. Next morning he headed for home. Barring a breakdown or bad luck, he made the round trip in one week.

Freighting was easier during the summer. The weather was warm, and there was grass for the horses. No hay had to be carried in the wagons, and no holes had to be chopped in the ice to water the horses. Summer rains were troublesome, and the mosquitoes were bad. Even so, freighting in summer was more pleasant than in the early spring when the roads were wet and muddy and the freighter faced getting stuck occasionally.

Dad continued to freight until 1919. The summer of that year he contracted to fence some land on Cottonwood Creek above Big Piney. The Standard Timber Company was building a commissary camp about twelve miles north of where Dad was working. He investigated the possibility of employment in the tie camp. It is probably fortunate that he decided to give up horse and wagon freighting. Increased hay prices following World War I climbed to twenty-five dollars a ton which adversely affected a freighter's income. Road conditions were also improving, and trucks of a better quality were being made. Two Big Piney merchants tried having their freight hauled by truck. By 1922 freighting by team was almost a thing of the past.

In the fall of 1919 Mother, my two brothers, and I remained in Marbleton where I was enrolled in the second grade of school. Dad started to work at the tie camp. An old hand at timber work went out with him and showed him how ties were made. Dad made nineteen ties the first day. The next day he decided to put himself to a real test. On an adjacent strip of timber worked a Swede named Victor. The latter was not as big as Dad, but he was strong and husky and had hacked ties for a number of years. Every time Dad heard a tree fall on Victor's strip, he stopped whatever he was doing and falled one. A tie hack never spoke of felling a tree; instead, he referred to it as falling a tree. Dad learned later that Victor, knowing that a greenhorn was working on the next strip, made it a point to fall a tree every time he heard Dad fall one. Dad was determined to work up every tree he had cut down before quitting work that night. By the end of the day he had made forty ties which was a good day's work for even a seasoned tie hack. He was convinced that he could earn a living in the timber. The following day Dad's limbs were stiff and sore, but he continued to make ties.

Dad built a log cabin in which he lived alone. Cooking his own meals in addition to doing a hard day's work in the timber was not pleasant, but it was more economical than boarding at the cookhouse for $1.70 a day. His meals were quite plain, and he missed having an occasional pie or cake. Some Swedish girls told him that if he would leave a can of fruit outside his cabin some time, they would make him a pie. Every day they reminded him of their offer and asked when he was going to set out a can of fruit. At last he took them up on their offer, and he left out a can of tomatoes. They gave him a bad time, of course, when he returned from work, asking what he expected them to do with a can of tomatoes. His response was "What's the matter? Don't you Swedes ever have tomato pie?"

After a couple of months passed, he came down to Marbleton to move us to the tie camp.

CHAPTER TWO

The Moline Camp

It was about the middle of November in 1919 when Dad arrived. Bill and I were wild with anticipation as the folks packed the wagon box for the move. It would mean that Dad would be home with us again. We were intrigued with the stories he told us about life in the timber, the tall trees, snow higher than our heads, and the Swedish customs observed in the camp. The sled's wagon box was loaded with a limited number of household items, two bedrolls, food to eat en route, a little hay for the horses, and a quarter of elk meat and a quarter of deer meat which Granddad had given us after his annual hunting trip. A canvas was spread over the wagon box, and we were ready to travel.

Granddad accompanied us on the trip so that he could take the team and sled back to Marbleton. He arrived at our house before daybreak the morning we left. After breakfast the team was hitched up, the house was locked, and we set out. It was with great excitement that we boys boarded the sled. The weather was cold. Mother, Bill, and Herschel sat in the wagon box under the canvas with bedding spread over their legs to keep them warm. Most of the time I sat on the wagon seat with Dad and Granddad. From time to time while one of the men drove the team, the other man and I would get out and walk behind the sled to warm our feet. As we plodded toward the mountains the snow became deeper. We stopped at noon, kindled a campfire, and had a quick meal. Freshly fallen snow was melted in a kettle over the fire to provide drinking water. I can still remember the horrible sagebrush taste of that water.

After finishing our scant meal, we started on again. We passed various ranches along Cottonwood Creek, including John Budd's, Nichols,' the Mickleson place, and Hake's. Toward the foot of the mountains the snow became deeper. There were fences along the way, and about seventeen gates had to be opened and closed. A gate always told us that we were still on the trail, and we must have passed them all by late afternoon. Toward evening the men became aware that we were not on the road, actually a sled trail over which freight was hauled weekly to the tie camp. Nothing was visible in any direction but snow and more snow. One man walked ahead of the horses in search of the trail and to make sure that the snow was firm enough that the horses would not get bogged down. With darkness almost upon us, it was decided that we had better stop for the night. The horses were fed and tied up, a campfire was started, and supper was prepared. We then unrolled the two beds in the wagon box and crawled in for the night.

There was not enough firewood to provide a fire the next morning for getting warm and cooking breakfast. Sagebrush, the usual fuel for campers on the prairie, was covered with several feet of snow. We had not yet reached the forest where pine wood was available. The folks had loaded an old bedstead in the wagon box. It did not take them long to decide that they could get along without it in the tie camp. One of the men chopped it up for firewood. We boys were given a quick breakfast, directed to warm our feet over the fire, and then required to get back under the canvas and to cover our legs with bedding. Soon the horses were hitched up, and we were on our way again. The men quickly found the trail in the daylight, and travel was a little faster.

We pulled into the commissary camp that afternoon. After getting some baled hay for the horses and picking up a few supplies at the commissary, we drove the remaining three miles to the Moline camp. As we approached the camp, Dad stopped and chatted with a tie hack who was piling his ties near the road. I can still remember the fragrance of the pine trees and the freshly hewn ties.

The first thing Dad did when we reached the cabin was to check his sourdough supply near the stove and ascertain that it had not frozen. He must have arranged for a neighbor to build a fire in the cabin because the latter was quite cozy.

The cabin was about twenty-four by fifteen feet in size. The logs were hewn inside and like other log houses of the day the cracks between the logs were chinked inside with pieces of wood. On the outside the cracks were filled or daubed with mud. A shed made of slabs was built onto the front of the cabin. The only door led into the shed which sheltered a number of provisions and a large supply of firewood. The snow which was already quite deep eventually reached a level of eight feet that winter. Snow had to be shoveled away from the windows and the door of the shed. Stairsteps were dug into the packed snow from the door up to the level of the snow.

Our furniture consisted of a table, two beds, a handmade chair, a bench, and a few wooden boxes on which to sit. There was a small range near the door and a little heating stove near one of the walls. This was to be our home until spring when we moved into a larger two room house on the other side of the road.

The Moline camp was populated primarily by unmarried men. One of the more notable was Andrew Bloom, a likeable, humorous old character who hauled ties. He seldom went anywhere outside the tie camp, made a good living, and as he had no family always had cash on hand. If a man or a family had financial troubles, it was always Andrew who sensed it first and came through with a loan or a gift. He was also generous to children. During our stay in the Moline camp, we three boys happened to be at Andrew's house one day. There was a pile of cash containing about five silver dollars and some small change on his bunk. Suddenly Andrew picked it up, jingled it in his hands for a moment, and then handed Herschel a silver dollar. The latter, aged three, declined it and said, "I already got some money."

Andrew questioned, "Are you sure? Have you got plenty."

Herschel replied, "Yes."

Andrew said, "Vell, you better take this." Herschel accepted the dollar. Andrew then handed a dollar to Bill and gave me the remainder of the cash. Trying to decline it was futile for he insisted that we accept it.

It took two cookhouses to accommodate the tie hacks in the Moline camp. Several of the tie hacks lived alone. Four families with children lived in the camp. In the Moline family there were two boys and two girls. The Gills had twin daughters. There were a boy and a girl in the Cranor family, and there were three boys in our family.

Even on days when it snowed, one could hear the squawking of jays and the chattering of pine squirrels everywhere around the camp. There were a few blue jays and flocks of mountain jays, commonly called camp-robbers. They were attracted by discarded food, and in a short time they became so tame that they would eat from a person's hand. The squirrels were easily tamed, also. I remember that at a later date two tie hacks, Nils Nilsen, a Dane, and Sam Peterson, a Swede, each had a pet squirrel. If either man heard a squirrel chattering, he would open his door and call, "Come, Pete!" The squirrel would approach him and eat the bite of bread offered it.

It snowed almost every day. Occasionally there would be a heavy storm or blizzard. The snow continued to get deeper and deeper. During severely stormy weather, we were kept inside except for trips to the outhouse. This really gave us a taste of cabin fever. On most days we were allowed to go out and play in

(Nils Edin, ca. 1935)

Winter scene in a tie camp

the snow. The children in our camp made snowmen, threw snowballs, or coasted on skis and sleds. Occasionally our feet got frostbitten. We learned that it helped to wear heavy woolen socks. Mother cut up an old sheepskin lined coat and made moccasins for us. We wore the woolen socks and moccasins inside our shoes. With these and rubber overshoes we managed to avoid severe frostbite.

Coleman gasoline lamps did not make their appearance in the camp until the following year. We relied on kerosene lamps for light. A great inconvenience occurred a month or so after we arrived at the Moline camp when the commissary ran out of kerosene. The folks had brought a few candles, but three or four candles did not provide much light and the supply was limited. We solved the problem by making candles. Three strands of twine were braided together; then we dipped the braided twine in melted tallow, lifted it out, held it outside the door for a few seconds, and dipped it again. By this process we produced quite a number of candles. The light provided was far from satisfactory, but it got us by. We were much relieved at the end of the next week when the teamster brought a supply of kerosene from the commissary.

Our most tragic and frightening experience occurred in January when an epidemic of diphtheria struck the camp. Christine Moline, aged twelve, died. During her illness warm bedding was packed around her, and she was taken down to the commissary camp. The doctor from Big Piney was summoned to attend her, but he did not arrive until the day after she died. All of our family had diphtheria except Dad. He remained home from work and took care of us, performing the gruesome and painful task of swabbing our sore and swollen throats and feeding us a foul-tasting chocolate syrup laced with quinine.

About March the commissary ran out of eggs, and we children had no eggs for Easter. On a stormy Easter Sunday, Mother gave a party indoors for the bunch of us. She provided us with egg-shaped pieces of homemade candy made from powdered sugar. We also had a quantity of fudge. How the cookhouses made out without eggs I do not know. To a Swedish tie hack, breakfast was not breakfast without hot cakes; and I cannot imagine a cook making hot cakes without eggs. Perhaps the cookhouses had a supply of storage eggs on hand. At any rate, it was a great relief to everyone when the teamster arrived with eggs along with our other provisions after several weeks.

For the remainder of the year I do not remember any other calamities. Life went along fairly smoothly. Dad had his breakfast before sunrise and headed for the timber regardless of the weather. Often his strip of timber was so far from home that he took his lunch with him rather than take time to come home for his noon meal.

We boys eagerly looked forward to his return in the late evening. He stopped outside the door, shook the snow from his jacket, and swept the snow from the rubber shoepacs he wore. His clothing was wet from working in the snow. Upon entering the cabin he changed into dry clothing, washed his face, and combed his dark brown wavy hair which he always parted in the middle.

Dad stayed home on Sunday. Even though he had to cut firewood and sharpen his tools, he managed to crowd in time to ski and coast with us boys and upon occasion he made an indoor toy for Herschel. Bill and I spent much time with our skis and sleds. Herschel was only two when we moved to the tie camp. He resented not being allowed to spend more time outdoors, and at one time he stated that when he became a man he would build a house and stay out of it all he wanted. For some time life in the timber retained a certain novelty for Bill and me. To Herschel, however, it was the only way of life he knew.

Occasionally Bill and Herschel would get into a squabble over the possession of a plaything. Being bigger, Bill would come out the winner unless I intervened. I thought it deplorable of Bill to take advantage of Herschel who was so much younger and smaller. When Bill and I had a fracas, it was another matter. It did not occur to me that I was as much older than Bill as he was from Herschel. Usually we got along fairly well together. Sometimes there was competition regarding the acquisition of a first. If I said that I saw the teamster coming with mail and supplies from the commissary, Bill would say "Yes, I saw him before you did." My reply would be "You did not. I saw him first." Bill would respond "You didn't either. I saw him first." The argument would sometimes go on and on. I am sure no one ever really won.

The Gill twins sometimes came to our cabin on stormy days. Mother might read to all of us. We played various games, including such card games as old maid and steal pile. On most days we played outside.

We enjoyed playing in the snow, but as winter dragged on we longed to see bare ground again. Winter was the time, Herschel said, "when the ground gets up high." We began to have visions as to what we might do when the snow melted. I thought it would be fun to go swimming in all that water. One girl planned to use a large washtub for a boat and go sailing. Her less optimistic mother feared that the houses would be washed away.

It never occurred to us that the spring thaw would be slow and gradual. Eventually the melting snow did swell the tributaries of Cottonwood Creek, and snow was replaced by mud everywhere. Supplies from the commissary were no longer delivered by sled but by wagons pulled by horses which plodded through the mire.

As summer approached, the ground became dryer and the weather was warm and pleasant. There was no letup in work for the tie hacks. Those who were not on the annual drive floating ties down the streams continued to work in the timber. The work was easier in summer, and they could make more ties.

In the late spring Dad went down to Marbleton, hitched a team of his horses which were being cared for by Granddad, and drove them up to the camp. The team and wagon provided transportation for visiting in the other camps, for fishing trips, and for the one trip the family made down to Big Piney during the summer.

An episode occurred that summer which I have never

forgotten. I learned not to feed oats to horses. Dad pastured the horses on the green grass near the house. He also kept a sack of oats in the shed. After working the team he gave them a very small can of oats. One afternoon Dad and Mother went away from home but left us boys there. Old Pet, one of the horses, came into the yard. Bill was a little frightened. Had it been Shorty, he would not have been concerned for we had both ridden Shorty a number of times. We had been warned that people had been kicked and injured by horses, and we had misgivings about Pet. Bill decided that the way to prevent Pet from hurting him was to feed her some oats. He poured out a can of oats for her. It worked. She did not hurt Bill. She seemed to like the oats so he poured out some more. By this time Shorty had come into the yard, and he wanted some oats, too. It did look interesting, so I poured out some more oats for the horses.

About then a little Swedish girl came over to our yard to see what was going on. During the previous winter Mother had taught us to be polite to the girl. While playing near our house, she had stepped off her sled into a snowdrift up to her waist. Sobbing, she said, "Tan't det out. Never det out. Yust hafta stay here. Never det out." We told her to cheer up because she would get out when spring came. Mother came out of the house angrier than a hornet. She said to Bill and me, "I'm going to break off a little green branch from that tree over there and use it on both of you kids if you don't have that little girl out of that snowdrift in two minutes." We had her out of the snowdrift in half a minute. Then came a little lecture on politeness. Mother told us that when that little girl came to our yard we were to be polite to her. Well, on that warm summer day we remembered. We politely got an empty ten-pound pail for the girl and let her help us feed oats to the horses. The three of us took turns pouring out oats, and the horses kept right on eating.

When Dad and Mother returned home, Dad saw what we had been doing; and he grabbed Bill and started spanking him. Bill howled loudly enough to raise the dead. Dad then grabbed me, and he laid it on hard. I raised the roof two inches higher than had Bill. Dad said, "Don't you know those horses may die?"

For once in my life I sassed my dad. I said, "I hope your damned horses do die." Such impudence was never tolerated and the word damned was used only by grown men, but I figured Dad had already hurt me as much as I could be hurt. I was wrong. The second tanning hurt much worse than the first.

Dad was up all night doctoring the team of horses. They recovered, and three weeks later they were as healthy as ever. Bill and I also no longer needed to eat our meals standing up. Dad had cured the horses, and he had also cured two boys of feeding oats to horses.

Three years later Granddad rode horseback up to the timber to visit us. He got off his mare, watered her, and told us boys to put her in the barn. He said he would feed her later. We tied her up in a stall, unsaddled her, and put a little baled hay in the manger for her. Later that afternoon Granddad said, "Well, I'd better put on my overshoes and go down and feed old Dixie."

Bill said, "You don't need to, Grandpa. We already fed her."

Granddad replied, "Oh, all right then," and he picked up a magazine. He looked up suddenly and questioned, "You didn't feed her any oats, did you?"

Dad answered for us. He said, "No, they don't feed oats to horses any more." He was right.

By the end of the summer most of the tie timber near the Moline camp had been cut. The company established the Bloom camp a mile farther to the north. One cookhouse and two bunkhouses were torn down, and the logs were hauled to the new location and reassembled. Our family moved to the new camp, and Mother ran the cookhouse while Dad continued to make ties. The only employees there were the tie hacks. The teamsters who hauled the ties from the timber to the creek banks continued to live at the Moline camp and kept their horses in the barn there. We remained at the Bloom camp two years.

CHAPTER THREE

Making Ties the Hard Way

Wyoming had much lodgepole pine timber which was of limited value except for use as railroad ties. Millions of these cross ties over which rails were laid were needed for the maintenance of the Union Pacific Railroad tracks. Road conditions were still crude, however, and there was no suitable equipment available for transporting logs to sawmills from forests which were many miles from even the smallest cow towns. The only practical way to utilize such timber was to have ties hewn in the forest and then to float them down the streams in the spring when the water level was high.

Falling, measuring, scoring, hewing, sawing, peeling, and sometimes parking went on daily for ten hours a day six days a week. Perhaps it was not as tedious as assembly-line work at which a laborer performs the same simple task over and over for an entire shift, but it was hard backbreaking work more strenuous than most occupations. Swinging an ax with the force required to chop ties was extremely fatiguing.

Some fellows seemed to have the knack for hewing

ties and other men did not. When the Cottonwood Creek tie camp opened in 1919, a group of fellows worked at cutting logs and hauling them to build the tie boom where Cottonwood Creek meets Green River. After that job was completed only two of the men decided to remain in the timber to try making ties. One stayed for the rest of the winter. The other found that he could make only three ties a day, and he returned to ranch work.

One sturdy ranchhand stopped at our house in 1924 en route to a section of timber to try hacking ties. He was accustomed to hard work and had a reputation as a man who would stick with a job once he had started it. A week later he stopped at our house again. He said to Dad, "George, I worked harder than I've ever worked in my life and six ties was the most I could make in a day." Dad said that he could stand up under hacking ties better than under pitching hay. The fellow said, "Not me. I can pitch hay with the best of them and it doesn't get me down, but I can't make ties."

A tie hack did not claim to be a skilled craftsman. If one of them professed to be a good carpenter or blacksmith, it would cause someone to smile and murmur "If he was any good at that kind of work, he wouldn't be up here swinging a broadax." Even so, hacking ties did demand a certain amount of skill. To pass inspection a tie had to be no thicker than eight inches and no thinner than seven inches, and it had to have a minimum of five inches of hewn surface on each side. An error in an ax stroke could ruin a tie and cause it to be rejected. Hacking ties, thus, did require considerable brawn and skill.

The first step in hacking ties was to fall a tree. After cutting off the limbs, the tie hack would lay his eight foot measuring pole on the tree and cut a notch every eight feet as far up the tree as he could still get an acceptable tie. Next came the scoring. Standing on the fallen tree, the tie hack would chop slashes into each side with a double-bitted ax. Cutting into the grain of the wood and chopping out all the large knots was necessary before he could hew the tree. While still

(Nils Edin, ca. 1935)

Hewing ties with a broadax

standing on the tree, he would then hew it on each side with a broadax. This tool had a sharp blade, was about twelve inches wide, and was flat on one side.

Scoring and hewing the top tie required the greatest precision. As it was the smallest tie, the man had to be especially careful not to get it too thin and still retain five inches of hewn surface. It was here that he often used a ruler which he carried in his pocket. Called simply a rule by the tie hack, it was either a small folding ruler or an eight inch one made from a board. He gently smoothed the sides of the top end of the smallest tie with his scoring ax rather than take a chance at not getting it the right size with his heavier broadax. Care had to be used not to wedge a tie. If it were thicker on the top side than on the bottom side, it would be rejected.

With the hewing completed, the tie hack sawed at every eight foot notch. Then came the peeling. All bark had to be peeled from the unhewn top and bottom sides. After the ties were completed, a man still could not collect any money for them until they had been parked in rows along a strip road. Each tie had to be hooked with a pickaroon and dragged to a pile. The strip road had to be cut clear of trees and windfalls. It was over these strip roads connected to a main road that teamsters hauled the ties on sleds during the winter to the banks of streams to await the spring tie drive.

Hacking ties in the winter was much more rugged than in summer. Forest Service regulations required that a tree stump be no higher than twelve inches. In winter the tie hack had to dig a hole around the tree through perhaps as much as eight feet of snow. The hole had to be about four feet square to give him room to undercut the tree with his ax and to saw down the tree. Peeling the bark from ties was more difficult in winter. Sometimes in the summer a man would make a quantity of ties and allow them to dry out before he parked them. They were a little lighter and easier to handle. In winter the ties had to be parked every night lest they should become buried in snow.

The tie hacks and the United States Forest Service officials were frequently at odds. A Forest Service man explained his position in this way, "If a man who had livestock has to cull his herd or get rid of some of his stock for any reason, he gets rid of the bum stock. He keeps the best stock. It's exactly the same way in forestry. We get rid of the poor timber and hold on to the best." The tie hack who had to cut rough timber had a different view. Timber to him was of no value except to be cut. His work was easier, and he could make more money by cutting good timber. He argued that conservation of timber was stupid because future technological development would render all the surplus timber of little value anyway.

The Forest Service enforced regulations regarding the height of tree stumps, and the tie hack was required to cut timber marked for cutting. He ran the risk of being penalized if he cut unmarked trees. Every tree marked for cutting by the Forest Service had to be cut down.

A skillful tie hack usually got the tree to fall right where he wanted it to fall. Sometimes a falling tree would lodge against other trees in the way, however, and this was always a headache to a tie hack. He might try to loosen the lodged tree with a peavey. If this did not work he might fall another tree against the first one. This would bring the first tree the rest of the way down unless the second tree failed to fall with sufficient force, in which case the tie hack then had two lodged trees on his hands.

The most usual and successful way to bring a lodged tree down was to fall the tree which was obstructing it. This required working under the lodged tree and was hazardous. It was imperative for the tie hack to get out of the way of the lodged tree as soon as the second tree started to fall. Falling trees was the riskiest part of the tie hack's work. There were many miles, often snow-covered, between the tie hack and a doctor. Occasionally a man was seriously injured or killed by a falling tree.

The average tie hack probably made about twenty--

five ties a day. Those who were more skilled and faster might make thirty a day. A few topnotch men made forty a day. It was possible for an outstanding broadax artist to make forty-five or more in a day but rarely did a man keep up this pace on a daily basis.

The number of ties a man could make in a day depended in part upon the quality of the timber. Some trees had exceedingly large limbs to be lopped off. Occasionally a tree would have huge masses of knots, called "catface" by the tie hacks. Such timber was difficult to work. Another problem was rot which was the bane of a tie hack's existence. A man might hew a tree and red sawdust would appear when he started sawing at the first eight foot mark. At least the first two ties were rotten, and nothing would be paid to the tie hack for the work he had already done on them. One of the bosses came through the timber periodically and inspected the parked ties. The boss carried with him a pot of paint made of lampblack and oil. A dab of paint on the end of a tie showed that it had been counted and credited to the tie hack's account. Each tie not rejected was classified as first-class or second--class, depending on the amount of hewn surface. More money was paid for first-class ties. The boss put a horizontal mark with a red crayon on each second-class tie. A cross made with the crayon meant that a tie had been rejected.

During the winter the ties were hauled from the timber to the creek banks. When there were several teamsters hauling ties in the same area, bells were attached to the harnesses of the horses. This enabled a teamster to know when another teamster was driving along the same road from the opposite direction. The jingling of sleigh bells could be heard during the day in the various camps and throughout certain sections of timber.

Sleds on which the ties were hauled consisted of two pairs of sled-runners, one in front and one in back. Over these was fitted a tie rack about eighteen feet long. A teamster drove along a strip road, stopped at each pile of ties, and using his pickaroon dragged each tie onto the rack. When the sled was loaded with ties,

(George Linn, 1920)

Breaking up the landing

he secured the load with log chains and drove to the creek bank where he unloaded the ties onto huge piles. These piles of ties were called a landing.

Some teamsters contracted to haul ties out of a given area, and others worked for a monthly wage. Often an outlying camp was designated by the name of the teamster who contracted to do the hauling. The camp where the chief teamster Ole Bergland lived was known, thus, as Ole's camp. There were also Moline's camp, Bloom's camp, and Nylander's camp.

The bosses during most of that era were Bror Edin, the camp manager, and John Wasburg, the timber boss. Edin was the chief boss of the entire tie camp. He supervised the commissary, hired the men, and had general charge of the operations in the various camps. The timber boss set out the strips of timber for each man to work and directed the work of the tie hacks. Both Edin and Wasburg inspected ties.

It was generally believed that the major qualification for a boss of lumberjacks was his ability to whip any man on the job. I am sure that was a myth. Even though they had sufficient physical strength, those men never relied on it to get a job done. The most successful bosses I knew were extremely patient and even-tempered men. They held their ground and remained firm in their decisions. If a man wanted more credit or to draw funds for booze while already in debt to the company, the boss would set limits and refuse to budge. A man might be irritated because the boss had not gotten around to taking up, that is, inspecting his ties or because he had been assigned to a rough strip of timber when he had just finished working a couple strips of bad timber. If so, such a tie hack did not hesitate to tell the boss off in strong language.

On one occasion when drunk Nils, a Danish tie hack, decided to air his grievances to the boss. Although physically a small man, Nils had good firm muscles and was a hard worker. He made many ties in a day and took pride in making good ones. He was kind and generous, but he almost never laughed or smiled. He seldom had much to say but could really speak up when he became angry. Aware of Nils' condition, the boss calmly ignored him. In order to get the boss's attention, Nils finally grasped the boss's nose and twisted it as he continued to cuss him out. The boss was big and powerful enough that he could have pulverized little Nils. Instead, he listened until Nils had let off sufficient steam and then went on his way.

If a man looking for a job wanted to learn to make ties, a boss would get an old hand to go out with him and show him how it was done. Upon occasion a man would pretend to be an experienced tie hack when he did not know the first thing about such work. A boss once came upon such a would-be tie hack trying to hew ties while standing beside the fallen tree. Another time he saw a man trying to hew righthanded with a lefthanded broadax, although it is necessary that the smooth side of the broadax be next to the wood. In each case the boss patiently showed the man how to make ties.

I first peeled ties for Dad when we were at the Moline camp during the summer I turned seven years old. A year later Bill and I sawed and peeled ties. We continued to do so during the summers. Still later when we were in high school or at the university, my brothers and I worked with Dad. We would fall the trees with a two-man saw. Bill and I did the scoring, and Dad hewed the ties. We boys did the sawing and peeling. By working together we were able to make eighty ties a day. I tried hewing from time to time but never got very good at it. We had to board and room away from home at a cost of twenty-five dollars a month each while we were in high school, so our summer earnings helped substantially toward meeting our expenses.

Dad was built for physically hard work, and he was an expert hewer. He could hack ties with the best of them. His height was just under six feet, and he did not look especially muscular. Although he weighed 172 pounds, he appeared slender. His physique was not passed on to his sons. Not one of us was very athletic. Bill probably came closest. Perhaps Herschel would have, but he sustained a skull fracture just before he was six years old. For a time his entire right side was paralyzed. After surgery and treatment he regained the use of his arm and still later learned to walk again. He remained rather awkward, however, and his muscles were not well-coordinated. While we older boys worked at various jobs in the timber for many years from time to time, I do not believe that any of us would have been as successful tie hacks as was Dad. James was heavier than the rest of us boys when we were teen--agers, but he was not very athletic. Our family moved to Laramie before he was old enough to help in the timber.

CHAPTER FOUR

A Bit of Scandinavia

Operation of a tie camp required the employment of a great many tie hacks and a number of teamsters. Many of them were Scandinavian immigrants. Most of the tie camp population was of Swedish origin, with a sprinkling of Norwegians, Danes, Finns, and others. Some had been in the United States for many years and were naturalized citizens, but there were always a few who were fresh from the old country.

There appeared to be little friction among the various nationalities represented in the tie camp, but different national groups did tend to cling together. If a Swede moved into a camp and found one bunkhouse was occupied primarily by Swedes, he lived in that bunkhouse. If a non-Swede came into camp, he moved into a bunkhouse housing native-born Americans or nationalities other than Swedes. The residents of one bunkhouse might be named Peterson, Olstrum, Ostland, and Nasland; and men named Edwards, Spencer, Baker, and Adams might be found in another bunkhouse. Old Nils Nilsen, one of the Danish minority, often lived in a non-Swedish bunkhouse, although he mingled with and was on friendly terms with most of the Swedes.

The Swedes were in a majority at the tie camp, and one usually heard more Swedish than English being spoken in the forest and in the bunkhouses. Many Swedish customs prevailed in various aspects of camp life. Among the reasons a Swede had for working in a tie camp was that he could relate to fellow workers who shared his culture and had similar tastes and interests.

I never heard a Swede express any deep appreciation for the beauty of his surroundings, the snow-capped peaks, the fragrance of the pines, and the valleys which were green in summer and covered with snow much of the year. Although he did not say so, perhaps absence from this environment for a prolonged period would have caused him to long to be back again. I did occasionally hear a Swede express disdain for crowds and congestion.

A Swede could usually be detected by his speech. If a listener incorrectly identified a person as a Swede, the latter was probably a Dane or a Norwegian. The affricate spelled with the letter "j" seemed almost impossible for a Scandinavian to master. Thus John was "Yawn," jam was "yam," and so forth. I remember seeing a cartoon showing a Swede saying "joosta be a yanitor." I never knew a Swede who would have said that. There are those who assume that a Swede would not only substitute the "y" sound for the "j" sound but also the "j" sound for the "y" sound. I never knew a Scandinavian who did so. I have known a few who would finally master the "j" sound but would say "jes" for yes and "junion" for union. Similarly, I knew a Swedish lady who would say home instead of "hoom," unlike her husband and many other Swedes, but this same lady also said "toles" for tools.

In my youth I was frequently accepted tentatively as a Swede. At various times a man would say to me "You ought to learn Svedish, kid. Then you could talk to your ma and pa."

I would quickly reply, "But my pa and ma aren't Swedish."

The response was invariably "But Linn! By golly,

that's a Svede name."

I would explain, "It is a Swede name, but the Swedes of that name spell it Lind. My name is spelled Linn."

Often the fellow would ponder a moment and say, "Vell, that's a Svede name yust the same!"

Occasionally one encountered a recent immigrant whose inability to communicate in English and unfamiliarity with American customs presented problems. Sometimes it appeared that the only English words in the newcomer's vocabulary were "Ya, shore!" Thus, if one said to him "Good morning," "It's a nice day," "How are you?" or "Where are you going?" the response was always "Ya, shore!" A story heard in the tie camps was that upon one occasion two fellows soon after landing in New York learned to say ham and eggs. They ate almost nothing but ham and eggs from New York to western Wyoming.

Regarding the way many of the Scandinavian immigrants butchered the King's English in the 1920s and 1930s, I should quickly state that I admire them for their ability to communicate in our language as well as they did. English is taught in Swedish schools to virtually all children today. Although Swedish immigration to the United States has declined greatly since World War II, the Swedes who do come speak English without the dialect of the lumberjacks of my youth.

While by and large there was little friction among the various nationalities represented in the tie camps, there were a few exceptions. A favorite cussword used by some Swedes was "Lapp javl," meaning Laplander devil. I assumed at one time that when certain fellows referred to others as Lapps, they were applying that label to Swedes from the far northern part of Sweden. This may or may not have been true. Since then I have been told that at one time many inhabitants of Skane, the southernmost province of Sweden, called all other Swedes "Lapps."

The Finns were the second largest national group represented in the tie camps. They tended to cling together. At one time on Cottonwood Creek one of the outlying camps was called the Finn camp. It was nearly always populated exclusively by Finns. Differences in background and language between Finns and other Scandinavians are far greater than I realized as a youngster. The Finnish language is not an Indo-European language. It is as different from Swedish as an American Indian language is from English.

Old George, one of the Finns, managed to communicate pretty well in English, although both his pronunciation and sentence structure were far from standard. In producing any word beginning with a blend, he omitted the initial consonant. Play thus became "lay," sleep was "leep," floor was "loor," and so forth. From time to time he liked to get "somet'ing to rink." "No vant to get runk," he would say, "yust to feel good." One afternoon he called on another fellow. George had a couple of "rinks," then a couple of more "rinks," and so on. Toward evening he said, "I no t'ink I go home tonight. I too runk."

The other man said, "Well, where are you going to sleep, George?"

He replied, "Oh, I leep on the loor."

After the tie hacks organized, they enforced a rule that no obviously intoxicated man should be allowed to attend a union meeting. George explained his absence from a meeting one afternoon in this way. "They say," he said, 'dons go to meeting today. You too runk.' I say 'Excuse myself'!"

On one occasion a dog barked and growled at George. The latter ordered, "Better you be still!" Later he told the dog's owner, jokingly of course, "Some day I gonna shoot your dog. And if you get in my vay, I shoot you, too. You got no lice on your dog anyvay." "Lice" was George's word for license.

A Finn camp or any camp in which a number of Finns lived could be distinguished by its bathhouse. A Finn living alone would also have his own bathhouse, consisting of a small log cabin equipped with a stone fireplace about three feet high. The fireplace protruded almost four feet into the room. There was no chimney, and the smoke found its way out through an open window. The area near the fireplace was black with soot.

A man would build a fire in the fireplace and let it burn completely out. He would then go into the bathhouse in the nude, throw a pail of water on the hot stones, sit on a small platform provided for that purpose, and get a good steambath. Generally there was a tub of water on the fireplace with which he sponged the perspiration from his body. It has been said that it was the practice of some of the older Finns to run out and roll in a snowdrift after completing their baths. Afterwards they went to their cabins or bunkhouses and whipped themselves with switches to restore their circulation. I never witnessed this latter practice.

The bathhouse afforded a person a convenient way of getting a bath. Some of the Swedes and other non--Finns took advantage of the bathhouse if one were available. One old Swede explained the superiority of the steambath by saying that ordinarily when one takes a bath he just rubs the dirt into the skin. By taking a sweat bath, he said, one gets clean inside the skin and out. I never fully accepted his argument, but all would agree that such a bath was preferable to getting one in a washtub in a crowded bunkhouse.

I saw only one non-white tie hack, a Negro surnamed Brown. It was reported that he had a brother who hacked ties elsewhere in the state, thus dispelling the myth that no one but a Nordic could hack ties. Brown spoke Swedish quite fluently. In fact, he called himself "the smoked Swede." At one time he was in a pool hall in Big Piney and overheard some Swedes speaking in their native language. From their conversation he gathered that they were new arrivals in America. He spoke to them in Swedish, and they were dumbfounded to hear a black man speak their language. Brown said, "I see my color surprises you. When you have been in this country as long as I have, you will be as black as I am." I have heard different versions of that story. It is probable that he pulled the same trick on more than one greenhorn in different parts of the state.

The only time I ever saw Brown was in the Moline camp when we were in school there. During the winter of that year the poor fellow froze to death. Some believed that he walked out of his cabin in his sleep. Others argued that if he stepped out into the snow the cold ordinarily would have awakened him; therefore, he must have been mentally deranged. Whatever the cause, he was found covered with snow outside his cabin.

Some of the Swedes had a flair for woodcarving. After his evening meal a man might take a piece of green pine wood and carve a fan, the figure of a bird, or an ornament of sorts. Such carvings often adorned the interior of a bunkhouse. Many hilarious Swedish folk songs were often sung by the tie hacks, especially after they had imbibed a few drinks. Some of the songs were not suitable for mixed company, but as Swedish-speaking ladies were not around on most such occasions the songs were sung resoundingly. It was said that Fred Erickson, the commissary camp cook at the time we moved to the tie camp, had been a traveling minstrel in the old country. In any case, he demonstrated considerable talent in Swedish folk dancing, singing, and composing songs dealing with real and imagined incidents of camp life.

As work in the forest became more modernized and as employees began to have more contact with the outside world, some Scandinavian customs and practices waned. This was especially true of the younger people as they grew older and of the later generations of Scandinavian families. The immigrants I knew in the early 1920s became as American as steak, turkey, and lemon pie; but they also remained as Swedish as hardtack, pickled herring, lute fisk, and ginger cookies. Many of us remember them and their customs with interest and respect.

CHAPTER FIVE

Long Winters and Lonely Evenings

A tie hack was assured of a roof over his head, a stove heated his bunkhouse or cabin, there was plenty of wood for the cutting only a short walking distance away, and water was available from a nearby spring. Prices of food, clothing, and tools were high; but one could always get the necessities from the commissary. Even though a tie hack was in the hole to the company, his credit was good so long as he continued to turn out ties.

Residents of the tie camps were not bothered by smog, congestion, or pervasive litter. For the most part a clean environment was taken for granted. The Forest Service did enforce regulations regarding the disposal of waste brush and garbage. Such regulations were generally appreciated by tie hacks, and some of them urged even stricter enforcement. When the spring thaw came and exposed a spot where garbage had been dumped all winter, the tie hacks wanted the garbage burned and all of the tin cans and litter covered up.

There were relatively few forms to be completed by employers and virtually none with which the employee had to struggle. Federal and state red tape was insignificant. A tie hack did not have to file an income tax return. He paid no union dues during the 1920s. The only funds withheld from his paycheck were for his board and purchases from the commissary.

Life was much less complex in the tie camps of the 1920s. Perhaps this is why a few years ago Mrs. Bror Edin, widow of the former camp manager, said to Mother, "Don't you sometimes wish we could turn back the calendar to the days when we felt sorry for ourselves for being stuck out in the sticks?"

There was another side of the story, however. Tie hacks worked ten hours a day and six days a week. If they worked fewer than sixty hours a week, they had a hard time making ends meet. It is true their housing was free, but it consisted of a log bunkhouse or cabin with crude and uncomfortable furniture. Firewood for the bunkhouse or cabin stove had to be cut on their own time, and it took much fuel to keep warm during the winter months. Water had to be lugged from the spring and heated in a pan or washtub on the stove for bathing or laundering purposes. After use the water had to be carried outside and emptied. During the winter the weather did not permit the quick drying of blankets. A man often had to sleep in his blankets for months without having them laundered.

Fuel for the Coleman lamp which furnished lighting cost fifty cents a gallon at the commissary, but in town it sold for from twenty-five to thirty-five cents a gallon. Groceries, or grub in the camp vernacular, and other necessities had to be purchased at the commissary. Other items needed by the men had to be purchased from the extremely limited selection carried by the commissary or from mail-order houses.

Sunday was a tie hack's day off. He had to take his turn cutting firewood for the bunkhouse and cookhouse. From time to time a fellow had to spend half of his Sunday sawing and splitting wood. Every week his ax had to be sharpened and his saw had to be filed. This was done on an evening after supper or on a Sunday. Bathing was done in a washtub on Saturday night or Sunday. It was also on Sunday that each man heated a pan of water over the fire, seated himself before a small folding mirror, lathered his face, and peeled off a week's growth of whiskers. Some men sent their laundry to town to be washed, knowing that it would be two weeks before it would be returned. If a fellow washed his own clothes, he had to find time to do so. With these additional chores a man sometimes knocked off work early on a weekday in order to get a good rest on Sunday.

It never occurred to us children that the routine followed by the tie hacks might get stale. Working sixty hours a week for week after week and spending almost all of their leisure hours in the bunkhouse seemed to be an accepted way of life. Their cramped living quarters caused little resentment. Each man had his bunk made of rough pine lumber or poles cut from the forest. There were no springs or mattresses. When a man moved in, he would put some hay or pine boughs on the bunk and over this would unroll his blankets and quilts.

After supper Gus might file and set his saw, and Victor might sharpen an ax. Olaf might mend a tear in his trousers. Nils and Eric would perhaps be seated on benches or crude wooden stools with a cribbage board between them, one of them counting "Fifteen two, fifteen four, and a pair is six, and a pair is eight." Walfred, Andrew, and others might chat in Swedish. Occasionally a group would play whist or pinochle in one of the bunkhouses. Once in awhile there might be a poker game, although there was seldom much cash to change hands.

All of the men read whenever there was anything available to read. They looked forward eagerly to the weekly arrival of the mail which always brought a fresh supply of reading matter. Quite a few subscribed for the **Svensk-American**, a Swedish language newspaper published in this country for Swedish immigrants. Some also received papers from Laramie, Salt Lake City, or Denver; and a few subscribed for magazines.

From time to time a few incidents occurred which may have indicated a degree of cabin fever. During the winter of 1920-21 I was not aware of any friction or quarreling among the inhabitants of the bunkhouse which housed most of the Swedes. Life in the other bunkhouse was a bit different, however. Sam, an Austrian, cussed out another fellow in the middle of the night because the man snored so loudly that Sam could not sleep. Several fist-fights occurred between Ed and Barney. Other men also grew irate with each other from time to time.

A tie hack did not go by the calendar as to when winter began nor did he consider March 20th the first day of spring. Winter began when the snow on the ground did not melt and soon became covered with more snow. The winter of 1920-21 was a long one. Snow continued to fall until the cookhouse in the Bloom camp was covered, and we had to shovel snow away from the windows and door.

Bill and I went up on the hill behind the cookhouse and slid down on our skis and over the top of the house. We lived in the cookhouse. Bill and Herschel were not yet of school age, and there was no school for me to attend. I spent my time reading. Bill and I learned to play casino. Christmas brought us a variety of toys and additional books for me to read.

Dad got up about five o'clock in the morning and started the fire. He then helped Mother prepare breakfast for the tie hacks. After breakfast he went to work in the timber. In the evening after the supper dishes were done and the table set for the morning, he was glad to crawl into bed and read a magazine. Mother was kept busy during the entire day except for a short rest in the afternoon.

When the spring drive was launched, all the men were ordered to work on the drive until the ties were all in the water and the drive was underway. Those who wished to do so were then allowed to return to work in the timber. Consisting of tents, the drive camp was located nearly a mile below the Bloom camp.

In late April of 1921 Dad came down with pneumonia. The Big Piney doctor made one trip out to the camp at a cost of one hundred dollars. Dad lay in bed

(George Linn, ca. 1935)

Drivers at work

constantly for a full month. Mother continued to run the cookhouse. One of the Swedes said to her, "Yust call for ve if you need us." Walfred and Victor walked back to the Bloom camp every night to see how Dad was getting along and to offer any help they could to Mother. They spent the night in the bunkhouse and returned to the drive camp early the next morning for another day's work. Dad recovered and returned to chopping ties in the early summer.

While our way of life was generally well-accepted, a little grumbling might be heard from time to time. Some of the fellows said a school for the children was taken for granted in several other tie camps in the state. The Fox Park tie camp near Laramie, it was reported, was almost like a little town. There was a store which was not company-owned, a dance floor, a pool table, and other means of recreation including an occasional movie.

Early in the summer of 1921, Al, Bill, and Bob showed up in the Bloom camp and moved into the non--Swedish bunkhouse. They soon aired their dissatisfaction with conditions in the camp. They had been working at a large lumbering operation on the west coast where they had comfortable living quarters with electricity and indoor plumbing. Al was a tall, lanky bald fellow. Bob was a large strapping man with a big bushy moustache. Bill was a short chubby man. The three addressed each other as "Mr. Al," "Mr. Bob," and "Mr. Bill." At the table in the cookhouse one would hear:

"Will you have some meat, Mr. Al?"

"Thank you, Mr. Bob."

"You're welcome, Mr. Al."

A little later you might hear "I'd like some bread, please, Mr. Bill."

"Certainly, Mr. Bob."

"Thank you, Mr. Bill."

"You're welcome, Mr. Bob."

They were friendly with the other fellows, both Scandinavians and non-Scandinavians alike. The Swedes accepted them pretty well but were a little overcome by their formality.

Al complained and found fault with the quality of the timber, the timber boss, the bunkhouse, the merchandise from the commissary, and virtually everything in the camp. Most of the men got a trifle tired of his griping.

Bill was far less belligerent about most of the conditions, but he whined considerably about the absence of running water in the bunkhouse. When a man works in the timber, he is bound to get pitch on his skin and clothing. After a short time the pitch turns black. Bill thought it was deplorable that a man should have to endure that blasted pitch, but he was the only man in camp who did nothing about it. The others managed to wash it from their bodies and clothing.

In the fall, Mr. Al, Mr. Bob, and Mr. Bill returned to the west coast. The other tie hacks probably envied them, but nobody in camp seemed sorry to see them go.

The Bloom camp was abandoned in the summer of 1922, and we moved to the Nylander camp. Most of the men lived in this camp or in other larger camps where there was more mixing. Several families lived in the various camps. Dances in one cookhouse or another were more frequent. Occasionally several couples and certain of the single men would get together and play cards on Saturday night. Bill and I learned to play cribbage that year. We played two-handed or occasionally four-handed with our parents or others. Several couples also played five hundred.

The first radio in the Cottonwood tie camp was purchased by Bror Edin in the fall of 1924. It was a battery-operated set and was installed in the living room at the commissary camp. Reception was usually nil during the daytime, but after supper several of the fellows would gather around and listen to programs from KOA in Denver, KSL in Salt Lake City, and other places. They marveled that even in a far-off tie camp they could get music and programs from such a distance with no wire in between. They were impressed with the clarity and intelligibility of the reception as compared with that of the phonographs of the day.

The old camp flunky found it hard to believe that

they did not have to put a record on the radio or even wind it up with a crank. When a station signed off for the night, the standard time was usually announced. All the fellows would immediately reach for their watches and reset them if necessary. The old flunky refused to do so. He was proud of his expensive watch and was not about to reset it for any radio man.

In March 1925 several people from the outlying camps gathered at the commissary camp where they spent the night before Calvin Coolidge's inauguration as President of the United States. They hoped to hear the ceremony next morning. The radio reception was good that night as they listened to reports of the plans for the big event. Next morning they all sat quietly and listened intently and patiently while Bror tried diligently to tune in on the inauguration. He got nothing. It was a big disappointment, but the tie hacks continued to enjoy the radio from time to time.

Grover, the bookkeeper, bought a radio and set it up in the commissary the following year. It was about six years before there were other radio sets in the tie camps. They were a source of much entertainment to the few who could afford them.

A great many tie hacks might spend the greater part of a year without any social life with anyone outside their camp. Occasionally a group would get together for a dance in the cookhouse in one camp or another. Such functions were more frequent during the later years than in the early 1920s. These events usually occurred on a Saturday night, a Sunday afternoon, or a holiday. Music was provided by a phonograph or sometimes by an accordion player. If a Swede had a phonograph he would have a collection of Swedish records, including hambo and schottische music. Much of the music and dancing was of the same variety found generally in other rural areas of western America.

We went to our first tie camp dance at Thanksgiving time shortly after arriving at the Moline camp. A bunch of us rode on a sled down to the commissary camp where the dance was held in the large cookhouse. Our supper was prepared by Fred Erickson, an excellent cook. Fred kept a table set up during the entire night and provided cookies, cake, and coffee. There were families from at least four different camps present, along with a few couples from ranches down the valley. It was not a large crowd. There were never more than five couples on the floor during the entire night.

We marveled at the way Fred and a few others swung and whirled over the rough pine floor to Swedish music. Dad remarked that it must be some kind of Swedish dance. Later we learned that it was the hambo. We children went to sleep while people were still dancing. They must have danced all night, for people were still on hand when we awoke the next morning. Before leaving for home we all had a hearty breakfast in the cookhouse.

We moved to the south fork of Cottonwood Creek in the late summer of 1924. We lived the next two years in a four room log house at the edge of the forest one mile from the Burns camp on the other side of the creek. The following year was less lonely. More people had serviceable automobiles. In the early fall some of us might drive down to Big Piney or Daniel for a Saturday night dance. Occasionally families from the various camps would get together. There were more frequent Saturday night card games. After the first severe snowstorm, however, we were in for the winter.

A tie hack might go "outside" for a short respite after a long hard winter. If he went all the way to the railroad on the annual drive, he would often take a few days off for a fling in a bootleg joint. For most men such sprees were usually brief. With the snow off the ground, a tie hack could make more money; and he would soon get back to the timber.

CHAPTER SIX

The Cookhouse

The sound of the cookhouse bell was always welcome in a tie camp. The bell was usually an iron triangle; but if one were unavailable when a cookhouse began operation, an old broadax blade or a discarded circular saw was substituted. The cook rang the first bell at six o'clock in the morning to alert the men that breakfast would be ready in half an hour. At the sound of the bell, men would come pouring out of the bunkhouses and file into the cookhouse. At the Bloom camp Walfred Peterson, Gus Satline, Swan Johnson, Andrew Ostlund, Victor Latine, and upwards of a dozen or more men would seat themselves at the long oilcloth--covered pine lumber tables.

Packages of corn flakes, bowls of oatmeal, platters of bacon and eggs, and stacks of hot cakes were on the tables. The cook continued to make hot cakes to replenish the supply throughout the meal. A tie hack frequently put away from six to eight hot cakes on work days, along with bacon, eggs, cereal, and fruit.

One of the tie hacks was a heavy husky fellow with blond hair and a full-grown blond moustache. He would put a quantity of applesauce on his plate, bury several buttered hot cakes in the applesauce, put some bacon and eggs on the side of his plate, and pour syrup over the applesauce and cakes. After eating it and another serving or two of applesauce and hot cakes, he would sit at the table, pick his teeth for a moment or two, and then leave. An indispensable item on the table was lump sugar, called "bit sugar" by the Swedes. One of them would put a cube of sugar in his mouth, sip some coffee, chew up the sugar, and swallow it.

The noon meal was served at twelve o'clock. Called dinner in the tie camp, it consisted of meat, potatoes, several kinds of vegetables, and homemade bread. For dessert there would be canned fruit, pie, cake, or pudding. Supper at six o'clock was similar to dinner, but some variety was provided. Dinner fare included steaks or chops; and roasts, boiled meat, or fish were served for supper.

Hungry Swedes could do justice to a steak, although some of them frankly admitted they preferred pork chops. Like the typical Americans with whom they worked, they did not care much for hash; but they enjoyed meatballs. Swedish hardtack was on the table each meal except breakfast. This commercially-prepared rye bread was baked in disks. Even though a Swede ate several slices of fresh homemade bread, he nearly always had a piece or two of hardtack. Many a person who had never before tried hardtack learned that with a little cheese it made a good sandwich. When spread with a mixture of honey and peanut butter, hardtack was really tasty. The Swedes seemed satisfied with American breads, cakes, and cookies. Although pie was a strictly American dessert, I never knew a Scandinavian who was not wild about any kind of pie. He also enjoyed having Swedish cookies and pastries.

Tie camp customs sometimes rubbed off onto the non-Scandinavian. Turkey was popular at Thanksgiving or Christmas time, but Christmas dinner was not complete for a Swede without his lute fisk (lye fish), and we children grew to be crazy about it. We would not have swapped it for turkey. Even though Mother had not ordered it, the commissary sent a supply of that hard, dry stock fish to the cookhouse in our second year in the Bloom camp some time before Christmas. To make the fish edible, it had to be soaked in lye water for about a week. It was then thoroughly rinsed, boiled, and served with a white sauce. Several fellows gave Mother advice as to how long to soak it, but none knew how much lye to put in the water. Mother finally had to ski a mile to the Moline camp to ask the cook there how much lye to use.

Pickled herring was considered a delicacy and was a staple item on the table. We Linns liked it, too. One time a Forest Service man stopped at the cookhouse for a cup of coffee. He had visited a Swede at the latter's cabin earlier that day, and he told Mother with considerable indignation that the man had actually tried to feed him some raw fish! Mother chuckled and told him that for some time he had been eating her pickled herring, which was indeed raw.

We children had been taught to say "Excuse me" before leaving the table if others were still eating. One time a Swede asked Mother why we did so. It was a little hard for him to understand why we should have to be excused from eating. In the old country, he said, the child said "Thank you very much." We later observed this to be true. There were very few immigrant children in the tie camp, but those who did come from the old country would take the hostess's hand after a meal, curtsy, and say in Swedish "Thank you very much."

As soon as the dishes were washed and dried, the tables were always set and remained set during the day. A place-setting consisted of an unbreakable white enamel plate turned upside down to keep the dust away from the inside, silverware, and an enamel cup turned upside down on the plate. The table also held catsup bottles, sugar bowls, cans of evaporated milk, salt and pepper shakers, plates of hard dry toast (zwieback) from Johnson's Bakery in Denver, and a plate or two of cookies, doughnuts, cinnamon rolls, or cake. A huge pot of boiled coffee was always on the kitchen range.

On one of his runs through the camp a teamster would invariably stop at the cookhouse for a cup of coffee and a piece of toast or some of the pastries. If a man were cutting wood for the cookhouse or had occasion to be in camp in midmorning or midafternoon, he would stop for a cup of coffee.

Like the other Swedes, William Olson had to have his coffee at midmorning and midafternoon as well as with his meals. Called simply "W" because he always signed his name "W. Olson," the commissary camp flunky firmly declined to eat ice cream which was considered a luxury and greatly enjoyed by the other fellows. "W" said it was no good. "It freeze you up," he said. On one warm summer afternoon when he came in for his coffee, "W" was offered some fresh cherries, a rare delicacy in a tie camp. "W" declined them. "I yust gonna have a little hot coffee," he said.

Old Ralph, a big French-Canadian, was a quiet, matter-of-fact hardworking man. He was civil to the others but did not socialize much. When he was drinking he became more talkative and seemed to take on a rather weird personality. After he had been hitting the bottle freely one afternoon, he went into the cookhouse and sat there. Mother became a little uneasy and wanted him to leave. The camp manager was away, so she located "W." She told the old flunky that Ralph was in the cookhouse and that she was a little afraid of him.

"W" said, "He is there now? I get him out." He went into the cookhouse and said, "Say, Ralph, you got no business in the cookhouse. You go now."

Ralph replied, "Who says so?"

"W" responded, "I do."

"Are you runnin' this camp now?" questioned Ralph.

"W" declared, "Yes, sir, I be. You go now." "W" was much older and about half the size of Ralph, but the latter got up and left.

There was little conversation during meals. About all one heard when the men were eating was the clicking of knives, forks, and spoons on the enamel dishes and perhaps an occasional "Pass the hot cakes, please." After a man had finished his meal, he would quietly get up, carry his dishes into the kitchen, and place them in a large dishpan or tub. He would then take an apple from a bowl on the table and a toothpick and return to his bunkhouse.

A stranger visiting the cookhouse for the first time might feel that the atmosphere was as cold as the weather outside. Nobody seems certain as to how to account for the stony silence during meals. Conversation may have been discouraged earlier because it was in the cookhouse that quarrels and even fights then erupted. Actually, the fellows were not unfriendly or uncongenial. Conversation never lagged elsewhere in the bunkhouse, the commissary, or wherever they might gather. If they had something to drink, hilarious and lusty singing could be heard in the bunkhouses; but such behavior never occurred at mealtime.

Scrapes and even an occasional brawl did occur in the bunkhouses. Ed and Barney were a couple of drifters who had come into the Piney country and worked for a time on a ranch. Because the pay was poor they came to the tie camp looking for work. Ed was a big fellow, with an especially untidy appearance. He almost never smiled, and he frequently sneered. When he grinned or laughed, it was to ridicule something or somebody. If anyone accidentally bumped him, stepped on his toe, or disagreed with him, he came forth with a torrent of filthy, abusive, and sometimes threatening language. Barney was also a husky man, with a cleaner appearance and a kinder disposition. When attacked or threatened, he would put up a good scrap.

After being shown how to make ties, Ed and Barney were assigned to neighboring strips of timber; but they decided to work together until they acquired some skill with the broadax. They got into an argument as to how they should proceed. Ed became angry and took after Barney with a pickaroon. Barney retaliated, proved to be better with his dukes, and won the fight. Each then began working on his own strip, and they seemed to be on friendly terms again. Soon, however, they had another scrap.

Ed had no watch and counted on Barney to yell at

him at noon to let him know that it was time to knock off for dinner. One time Barney failed to do so, went to the bunkhouse, and washed up for the noon meal. With a sour expression on his face, Ed came in and walked slowly toward Barney. Barney hit him, and a fist-fight ensued. Barney again emerged the victor. Later someone asked Barney why he had hit Ed. Barney replied, "Oh, just because of the way he looked at me."

When the cookhouse bell rang, Barney went to dinner with the rest of the men. Ed remained behind in the bunkhouse to wash his face and doctor his bruises and skinned places. He then came down to the cookhouse. As usual, all the men ate in silence. Ed took the afternoon off and walked four miles to the commissary to buy a watch. Several other bunkhouse scraps between the two took place during the winter. A few quarrels among other inhabitants of the same bunkhouse also occurred. Such quarrels were never mentioned when the men were eating in the cookhouse, however.

I did see one fracas in a cookhouse. It had its origin in the Swedes' bunkhouse before mealtime. It occurred when old Gus and young Gus got into it. There were usually several Swedes named Gus in a camp, incidentally. It happened on a Sunday afternoon in the fall of 1921, the second year that Mother ran the Bloom camp cookhouse. Uncle Jerry, my dad's brother, had arrived to visit us. He asked whether the fellows gave us any trouble when they were drinking. Mother assured him that they did not. She said that when the men went on a binge they usually left camp. In any case, she told him, they always conducted themselves as gentlemen while in the cookhouse. Events later that evening were to cause her considerable embarrassment.

A few of the men had obtained some bootleg whiskey that day. Old Gus got his tongue pretty well oiled and began razzing the younger generation. He was an older fellow, perhaps about sixty, with light hair and a light graying moustache. For a number of years he had alternated between ranch work and timber work. His bowed legs indicated that he had been a cowboy. Much of the time he spoke Swedish with the others, and his English speech had a definite Swedish dialect. Even so, he began ridiculing the others for conversing in Swedish and for their poor English. Young Gus was a husky Swede who was hardly a greenhorn. He had served a hitch in the United States Navy. Some of the younger fellows resented old Gus's remarks, and at last the two Gusses came to blows. Old Gus sustained a bloody nose and came down to the cookhouse, asking for water to wash his face. He could not wash in the bunkhouse because the sober fellows had thrown out the belligerents.

During supper Andrew came into the cookhouse. He was the chief teamster for the camp although he lived and kept his team in the Moline camp. He had been visiting in the bunkhouse earlier and imbibing with the others. As he came in, the older Gus said, "You can sit down here by me, Andrew." Andrew misunderstood, thinking that Gus had said, "You can't sit down here." He threatened to drag Gus outside, tear him into pieces, and scatter him around in the willows. Probably he would have tried to do so if Dad had not stopped him.

At last Andrew was persuaded to sit at the vacant place beside the older Gus and have his supper. As they were eating, a couple of the fellows spoke briefly in Swedish. Old Gus asked them why they did not speak English and why they must act like greenhorns. Gus muttered something in Swedish later. Andrew said, "Vy, Gus, that is Svedish. Tomorrow I vill take you down to the schoolteacher. She vill learn you to talk English." Still smarting from the encounter with the younger Gus, the older Gus sat leaning across the table with his head in his hands.

In a conciliatory effort Andrew urged him to eat. Getting no response, he said, "Vant some coffee?" When Gus still did not reply, Andrew lifted Gus's cup and poured the coffee on the latter's head. Old Gus packed his suitcase and left the camp the next morning. Fortunately, such incidents were not commonplace.

Running a cookhouse was a tightrope operation. The company sent a statement to the cookhouse at the end of each month just as it did to each employee. The statement showed the amount charged against the cookhouse for grub and supplies. It showed on the credit side the amount charged each boarder according to the number of meals he had eaten during the month. If the cookhouse ended up too far in the hole, the cook was in trouble with the company. On the other hand, a man who worked ten hours a day swinging an ax or loading and unloading ties and who paid $1.70 a day for his meals expected satisfaction in both quality and quantity.

A good cook carefully observed what the men liked and did not like. It was essential to have sufficient grub on the table, and it was safer to overestimate than to underestimate. Sometimes there was food left over. Wasting food had to be avoided, and this required serving it in another meal in palatable form. The cook usually figured out how to do so, and such "warmed overs" supplemented the fare of a later meal.

Generally a tie hack expected to pay for three good meals a day. Much to the exasperation of the cook, however, there would be a case of mooching once in awhile. A fellow would get to drinking heavily in the afternoon, miss his supper, sleep off his drunk, and miss his breakfast the following day. At about ten o'clock he would show up at the cookhouse and fill up on coffee, toast, and rolls or pastry. He would then be the first one at the table at dinner time and take on an especially big meal. For two days, thus, he would be charged for only four meals and yet eat as much as a man who was charged for six meals.

While most employees might be fortunate enough to get an occasional day off, the cook rarely did. The man or woman running the cookhouse took time off only for a very important reason. Arrangements then had to be made for a substitute. The cook was the first one up in the morning and the last one to finish work at night. After washing and drying the dinner dishes, setting the table, and getting the meat cut and in the pressure cooker for supper, the cook might get a few hours rest in the afternoon.

Such rest periods, however, were necessarily limited because the cook had to crowd in time to scrub floors, peel vegetables, bake bread, and perform many other tasks incidental to running a cookhouse. Meat was furnished by the commissary by a quarter of an animal. The cook had to cut it up; and if hamburger, meatloaf, or Swedish meatballs were on the menu, it had to be ground. It was important that the cook keep a close check on grub and supplies. To run out of an item meant having to go without it until the next delivery from the commissary the following Sunday.

When the number of boarders exceeded twenty-five or so, a second cook or flunky was sometimes employed. During the later years of the tie camps, a flunky was often employed when there were even fewer men. His job was to wash dishes, set the tables, function as a waiter, and do various odd jobs. Even with his help, cooking for tie hacks was a rugged seven-day--a-week job.

CHAPTER SEVEN

The Commissary

When we moved to the Cottonwood tie camp in 1919 I soon learned that what was called the commissary was a store of sorts. I wondered why it was called a commissary and was told that it was because the company owned it. The commissary was housed in the only two story building and was the largest structure in the entire tie camp. Like all the other buildings, it was constructed of logs. It did resemble stores I had seen but with certain differences. There were no showcases, no advertising signs, and no displays of merchandise.

There were two counters. Behind and underneath one counter were items of work clothing for men. There were no clothes for women and children. Behind the other counter and in the spacious storeroom were the groceries, tools, and hardware. During the winter, meat was kept in another building behind the commissary. In the summer it was kept in a screened meathouse high off the ground where it was cooled by the breeze. At one end of the commissary was the office where the camp manager and the bookkeeper had their desks and mail was sorted. There was no post office in the tie camp, and mail was routed to Big Piney thirty miles away. Everyone looked forward to getting the mail which ordinarily we received only on Sunday. If a man went to the commissary during the week and there was any mail there, the unforgiveable sin was not to bring at least the first-class mail to everyone in camp.

A man's supply of gloves and socks had to be replenished frequently. Other clothing and tools had to be replaced from time to time, and prices at the commissary were high. If a man took time off for a rest or to take care of personal matters, he still had to pay board if he remained in the camp. Very infrequently a tie hack might draw his wages and go to town for medical reasons, to do some shopping, or for a rest and change. Upon his return he might then run into some particularly rough timber or get a poor inspection on his ties. In that event he was in danger of ending up in debt to the company.

On the wall near the entrance to the commissary was a bulletin board on which were posted the specifications for ties and the rates paid for first-class and second-class ties. If a man were lucky his ties might average twenty-five cents each during the early 1920s. Thus for twenty ties he would receive five dollars. Out of his wages a man had to pay $1.70 per day for board at the cookhouse. Instead of a cash register there was a small cash drawer under one of the counters. It contained little money and was seldom used. Purchases were recorded in chronological order with the customer's name in a book which was kept open on a counter.

The commissary was open seven days a week, but the busiest day was Sunday. If Eric Johnson broke an ax handle and had to get a new one during the week, he walked or skied to the commissary and picked up a new handle along with perhaps a few boxes of Copenhagen Snuff, called snoose by the Swedes. Most merchandise, however, was delivered by teamsters on Sunday. During most of the year, a teamster called every Saturday at the cookhouse for the weekly order of grub and supplies, orders from the employees, and outgoing

(Nils Edin, ca. 1921)

The Cottonwood Creek Commissary Camp

mail. Orders were filled on Sunday morning by Frank, the bookkeeper, or by Grover, who later succeeded him. The bookkeeper was assisted by old "W," the commissary camp flunky.

When I first knew "W" in 1920, he was probably about sixty. He was not a big powerful man, but he was husky. He had made ties earlier, and he did so again when his services were wanted as a tie hack. The company needed a jack-of-all-tasks around camp, and "W" filled the bill. He always presented a neat appearance. Except for his moustache he never seemed to have a growth of whiskers, and he probably shaved more frequently than did most of the employees.

Old "W" helped unload and weigh the freight when it arrived from town, loaded the sleds with merchandise for the camps, cut wood for the commissary and cookhouse, and waited on customers when the bookkeeper was out or busy with other work. He took care of a team of horses and during the summer delivered merchandise with a spring wagon. If an important message had to be delivered or an errand run to one of the outlying camps, "W" hitched up the team and made the trip. Most of the time he responded quite cheerfully when he was called upon. There were exceptions, however, if the pressure became too great. At one time he was requested to make a run with the team on very short notice. Old "W" exploded, "Oh, damn it to hell! I yust took the harness off them horses."

A young child could wander around the camp quite safely as "W" would look out for him. When freight was being unloaded and the youngster was in the commissary, "W" placed the child in a huge woodbox where he would be out of the way and safe from harm. The barn door was kept closed except when "W" was at work there. In that case if a child were around, "W" would put him in the manger until he was ready to leave. There were usually a few stray cats in camp, and "W" would pick up some scraps from the cookhouse and feed them.

Each summer when the drive was underway and there was little activity in camp, "W" would go down to Salt

Lake City for a few days. At one time he asked Mother and Mrs. Edin how much they paid for a room when they were in Salt Lake City. Their answer caused him to blow sky-high. When he was in Salt Lake City, he managed to find a room for seventy-five cents a day where he could cook his own meals. He went into a restaurant on one occasion and ordered a cup of coffee and a piece of cake. He was outraged because he was charged twenty-five cents.

"W" spoke English with a heavy dialect, and he also stuttered which made him hard to understand. Nevertheless, he was persistent and eventually always got his message across. He had obviously struggled hard to learn English and had acquired some of the rules for spelling English phonemes. If he had to deliver a carton to Dad and the bookkeeper had written "George Linn" on it, "W" would draw a line through it and write "Yords Lind." That was the name by which both Dad and I were known among the Swedes for a number of years. When waiting on customers, "W" carefully copied the labels from cans in order to spell the names of items correctly in the sales book. I never did learn how he managed to spell names of articles which were not in cans and had no labels.

On one occasion a load of freight arrived for the commissary when Bror Edin was away from the camp and no bookkeeper was on hand. With "W" to assist her, Mrs. Edin stocked the shelves with the newly--arrived merchandise. Over the protests of "W," Mrs. Edin placed many articles on the shelves in full view. "W" argued that she should "put 'em under the counter out of sight. This vay," he said, "everybody who come in see it and want to buy some; pretty soon there von't be none left."

On a typical Sunday morning the commissary would perhaps receive an order from the Nylander camp cookhouse for a case of eggs, a quarter of beef, a case of evaporated milk, and a quantity of canned peas, corn, or string beans. An order from Charlie Olson might include a saw wedge, a half-dozen saw files, five pairs of canvas gloves, and six cans of tobacco. Carl Anderson who lived alone might want a ham, a dozen cans of pork and beans, a package of oatmeal, two pairs of socks, and a supply of snoose.

No cash changed hands as purchases were written down in the book. At the end of the month each man got his statement showing how much had been credited to his account in wages and how much had been debited for meals at the cookhouse and for merchandise purchased from the commissary. He never knew with certainty the price of an item until he received his statement. The statement told him whether he was in the hole or had a credit balance. Any time he had money coming he could request a check, and it would be mailed from the company's Evanston office.

Merchandise from the commissary was delivered by sled in the wintertime. The sled used for such purposes was usually a small one known as a go-devil, consisting of one pair of sled runners under a box. Andrew Bloom and certain of the other Swedes called it a "go-dabbin." Usually at about noon on Sunday a teamster arrived at a camp with the merchandise and mail. We children shrieked with delight when we heard the sleigh bells as Andrew Bloom or John Moline drove down the hill and into the camp.

We liked to help unload the sled. There was usually a sack of candy for us, and we would get to read the funnies in the Sunday paper. The fact that the paper was a week-old one did not bother any of us. I read "Bringing Up Father," "Mutt and Jeff," "The Gumps," and "The Katzenjammer Kids." "Gasoline Alley," "Happy Hooligan," "Mr. and Mrs.," and others were added later. The first year in the Bloom camp I read them aloud to Bill who had not as yet started to school. If I had finished reading the story books I had and the children's magazines we received, I read the funnies again.

With freight arriving at the commissary only once a week and sometimes not that often, the commissary would run out of certain items. Frequent grumbling from the men could be heard when they had ordered something and did not receive it. A man might order a pair of leather shoestrings for his shoepacs and get one shoestring, or a fellow who lived alone might order oatmeal and get shredded wheat biscuit instead. Very rarely, though, did the management allow the commissary to run out of Copenhagen Snuff.

CHAPTER EIGHT

School in a Log Cabin

I started to school in the first grade at Marbleton in the fall of 1918. It was a one room school housing about twenty-five children in grades one through six. Pupils in higher grades walked a mile south to Big Piney. School was closed for a time that year because of the big flu epidemic. I was just recovering from the flu when news of the Armistice reached Marbleton. The following September I entered second grade there.

One concern that my parents weighed carefully before deciding to move to the tie camp in November 1919 was the absence of any school. In addition to the children in our camp, there were a few other camps. It was planned that a school would be started in the spring. The school year would run through the summer when travel was easier and would extend through the fall and early winter. We could thus complete a nine month school year.

Leona Linn, my aunt, taught the village school in Marbleton, and she sent along a few books for me to study during the winter. I spent a part of my time each day reading from second grade readers. Seated on a block of wood at a large chair which served as a desk, I also practiced writing the letters of the alphabet and doing simple arithmetic exercises.

Previously made plans for summer school did not materialize when spring came. The tie camp was in the same school district as the ranch lands in the valley below. I do not know whether failure to provide us a school was due to inability to find a teacher or to a political situation of which I was unaware. For another full year there was no school, and I continued to study at home. My parents purchased a small blackboard, and for Christmas I got a number of books including **Boy Blue and His Friends, Billy Whiskers,** and **Aesop's Fables.**

Often I would have preferred to read silently from my readers and story books, but Bill liked me to read aloud to him. Mother encouraged me to do so as it gave her a chance to check on my reading. If I read aloud "The wolf run into the forest," Mother would smilingly say, "Was it run or ran?" If I read "The goat come over the bridge," she would correct me by asking, "Was it come or came?" Later I might read "The girl brought the basket." The listening Bill would smirk and say, "Was it brought or brought?"

During the winter of 1920-21 I read much of **The Adventures of Huckleberry Finn.** Strangely enough, I believe that the two factors motivating me to read that book were funny papers and Mother's attempt to discourage me. I had become slightly acquainted with the characters of Tom Sawyer and Huck the previous year while we were in the Moline camp. The Gill twins brought their story books over to our cabin. Along with the children's version of **Robinson Crusoe** and other books, they brought a copy of **Tom Sawyer.** Mother read portions of it to us, and we all enjoyed the antics of Tom and Huck.

The following year Mother's relatives in Kansas occasionally sent her copies of the Sunday issue of a newspaper. I have no idea where the latter was published. One of the comics which ran in that paper was "Tom Sawyer and Huck Finn." The episodes in which they figured were partly the product of the cartoonist's imagination but included the characters of Mark Twain's novels. I recall that they had a woman schoolteacher, whereas in **Tom Sawyer** there was a schoolmaster. Even though they were a pretty cheap imitation, those funnies interested me in Huck Finn. I was pleased to learn that we had a copy of **Huckleberry Finn.** My mother told me that it was written for older people and would be too difficult for me to read. I attempted it, nevertheless; and after noting that I could read it, albeit slowly and laboriously, Mother encouraged me to continue. The farther I got with it the more it interested me.

Archie Gibson, a forester, was stationed by the United States Forest Service at the Moline camp the following summer. He was newly married, and the folks were pleased to learn that his wife was a teacher who would be glad to have a job. Mrs. Gibson was a college graduate with seven years of teaching experience. Her qualifications were far superior to those of most rural schoolteachers in western Wyoming at that time.

The school board granted us a six-month school. Dad and Swan Nylander quickly converted a cabin in the Moline camp into a schoolhouse. At the new teacher's direction they built a table to serve as a teacher's desk and a sand table. They made chairs for the children. Mr. Gibson nailed together some planed boards and painted them with lampblack to serve as a blackboard. Later in the year the school board provided us with conventional seats and desks, but at the opening of the year one desk was improvised by laying a Forest Service sign across a block of wood. Mr. Nylander and Dad agreed to cut the firewood for the year.

There were few books provided by the school district at the opening of school, but more came later. Mrs. Gibson wanted a globe, so she had her mother send her a small one from Boston. The former had to do a tremendous amount of improvising at the beginning of the year, but school did open.

After nearly two years in the tie camp, Bill and I set out on foot through the forest for school at the Moline

(George Linn, 1932)

The younger generation on Smith's Fork

First row: (left to right) Clyde Davis, Jessie Daniels, Jay Roop, Dal Johnston, Jimmy Linn, Sylvia Daniels, Margit Edin, and Nils Edin

Second row: (left to right) Ara Roop, Eddie Roop, Mrs. Bror Edin

Third row: (left to right) Mrs. Frank Edin, Johnny Edin, Mrs. Joe Johnston and son Stanford, Joe Johnston, and Mrs. Burton, teacher.

camp a mile away. We carried our lunch pail, tablets, pencils, books, and so forth. Nearly all of the children who had lived in the Moline camp the previous two years had moved away. Andrew still lived there, and Clyde and Marion, new arrivals, also lived in the Moline camp. Margaret had to travel a mile-and-a-half from the Nylander camp. Including us Linns, the tie camp school had an enrollment of six.

I do not recall a single absence due to bad weather. The road over which Bill and I traveled was sheltered by the forest. Margaret had to travel a greater distance, a part of which was through an open unsheltered area. If at the end of the school day a severe storm was raging, she sometimes spent the night with the Gibsons. I am sure that the Nylanders appreciated this arrangement, but they probably sometimes had considerable anxiety over not knowing with certainty whether the little girl had remained at the Moline camp for the night or had tried to make her way home through the storm.

Recess time was spent coasting down the hill on skis and sleds. It turned out to be a very successful school year. The small enrollment facilitated much individual instruction, and Mrs. Gibson proved to be an excellent teacher. We did not know then how fortunate we were. The six-month school year ended in March, Mr. Gibson was transferred to Afton, Wyoming, in May, and that was the last school we were to attend for more than a year.

Mother, my two brothers, and I took a trip to Idaho that summer to visit Mother's brother and his family. When we returned to the tie camp in early August, Dad had already started the ball rolling to get us a school for the coming year. I do not know what went wrong, but he was unsuccessful. We lived at the Nylander camp during 1922-23 without being in school. Bill and I continued to study at home.

When the summer of 1923 rolled around, there were three families in the tie camp who had children of school age. They all began contacting school board members and pressuring them for a school, but the board discouraged them. There was a long-established rural school at the Nichols' ranch nine miles below the commissary, and the school board urged the tie camp parents to send their children there. At last the other two families agreed to pay board and room for their children at the ranch owned by Billy Nichols who hauled the freight from Big Piney to the tie camp.

Nichols had purchased a log house at the edge of the timber and wanted it moved to his place. He told Dad that if he would move it for him, Mother, my two brothers, and I could live there during the school year. Dad hired a man to help him, and they moved the house down to the ranch. Dad then moved the family but returned to work in the timber himself.

All was ready for the opening of school. Alas, when the opening day arrived, no teacher showed up. After more than a month of no school and many complaints from patrons, a board member telephoned Mrs. Nichols and told her that a teacher had been hired. The teacher, he said, was at the home of another board member many miles away. If we wanted her we would have to go get her. When Mrs. Nichols protested that Mr. Nichols was away with his truck after a load of freight and that we had no way of going to get the teacher, the board member said, "Well, take her or leave her! It's your last chance."

Mother telephoned to the tie camp commissary and asked that some one go out and explain the situation to Dad who was working about two miles from the commissary. Dad's old Chalmers, his first car, was broken down, and he hiked about seven miles from the timber to the home of a rancher who had a Ford. The rancher had to work during the day, so Dad hired him to drive him at night to where the teacher was staying. They returned to the Nichols' ranch early the next morning with the teacher.

Miss Waldon (not her real name) said that she was from Oklahoma and had taught school seven years. She was eager to start school and did so that very morning. She took an avid interest in the ranch, remarked about the queer little leaves on the trees and bushes, and asked whether there were trout in the creek that ran through the property. She said that she had never seen a trout. At the supper table she was told that in the winter she would probably want to learn to ski. Gesturing with a knife and fork which she held side by side, she asked, "When you ski do you take steps like this, or do you slide each ski forward like this?" She also asked whether she might be allowed to ride a horse. She said that she had ridden Shetland ponies and Indian ponies but had never ridden a horse.

People soon concluded that Miss Waldon was rather strange. Crossing their fingers, they hoped that she would prove to be a successful teacher. It was revealed a few days later, however, that Miss Waldon was from a family who lived on an isolated ranch some distance from Big Piney. After being identified, the woman confessed that she had no teaching credentials, had never taught, and had applied for and accepted the teaching job on a bet. Her actions did not evoke much laughter on the part of our parents.

A few weeks after she left, a new teacher arrived and school started again. He was a nineteen-year-old man from Missouri. The hard winter and rural Wyoming life did not appeal to him, and he became lonely and homesick. Country dances did not interest him, and he longed for the type of socials to which he was accustomed where such games as hide the thimble were played. After about four months he resigned and returned to Missouri.

Finding a teacher with the school year more than half over was, of course, not easy. The school board finally succeeded in hiring an experienced teacher. By having school on Saturdays and allowing the school day to run until five o'clock daily, we had the equivalent of eight months of school by late June.

Life on the ranch was interesting and different from that we had experienced in the tie camps during the previous four years. There was still enough snow for skiing and coasting. In addition, there was also horseback riding and a taste of western ranch life.

The ranchers had telephones. Each telephone had its

own ring. Nichols' number was a long and a short and a long and a short. Ringing was done by turning a crank at the side of the telephone. To reach another phone, one rang a long and two shorts, and so forth. One long ring reached the operator, or central, in Big Piney. In order to reach a telephone outside our own area, one called central to get connected. This party line provided ranchers with a way of getting local news. If one heard another subscriber's ring, he put the receiver of his own telephone to his ear, held his hand over the transmitter, and listened. It was by such "rubbering" that announcements of dances in the vicinity were circulated.

I still remember vividly the country dances held from time to time on one ranch or another. Music was usually provided by a fiddle and a guitar. Someone would pass the hat to collect money to pay the musicians. Ranchers, cowpokes, and others would dance the waltz, the fox-trot, the schottische, polkas, and a number of square dances.

Billy Nichols usually called the square dances. Here is one of the calls he shouted out, as I recall it, heard above the fiddle and guitar:

Honor to the left and honor to the right,
Eight hands up and swing to the left and then
  to the right,
All go up and all come down,
Allemande left and promenade all.
Swing 'em in a corner like a'swingin' on a gate,
Chase that rabbit, chase that squirrel, chase
  that pretty girl around the butterfly whirl,
The ladies bow and the gents bow-wow.

Supper was served at midnight. The host always provided the coffee, and ladies from the various ranches brought salads, sandwiches, cakes, pies, and so forth. After supper the dancing resumed and continued to perhaps three o'clock in the morning or later.

The following year we had school at the tie camp for a full nine months. Gus Gill returned to work for the

Standard Timber Company the summer of 1924, and his twin daughters would be attending school in the eighth grade. The school board hired a teacher. Dad and Gus built a log cabin at the Burns camp on the south fork of Cottonwood Creek which would be the schoolhouse for two years.

We lived in a four room house on the other side of the creek. The Gills lived a mile or so from school in the other direction. We were joined later by the Nelson boys who lived on Bear Creek a mile away. This gave the school an enrollment of seven. We children sawed and split the firewood. We skied to school during the winter. Our parents bought us a long toboggan that year. Skiing down a steep hill near the school or coasting on the toboggan was a real thrill. Even though we were isolated from the outside world, life was not really so bad.

Our lives were much the same the following year. The Gill twins and the Nelson brothers had gone. Wallace McCready lived in a cabin on our side of the creek with his widowed father and his older brother. The schoolhouse was moved over to our area so that we did not have to hike or ski a mile to school and eat cold lunches at noon. Leona Linn Pape, my aunt, was employed as our teacher. Her husband, Paul Pape, worked as a teamster for the Standard Timber Company for the year she taught school in the tie camp. As during previous years, recess activities usually consisted of skiing and coasting.

After school we had to saw and split wood for the fires at home. Time did not hang heavily on our hands. On weekends we boys frequently loaded some bacon, eggs, bread, potatoes, a frying pan or two, and some unbreakable dishes onto the toboggan. Pulling the latter, we then trudged to the top of the steep hill between our house and the creek. As we coasted down the hill, we aimed the toboggan toward a clump of pine trees in the valley where for some reason the snow was a little shallower. Sometimes we sailed through a raging blizzard.

Upon reaching the grove of trees, we hastily kindled

a fire with dry branches. Often our hands were so numb from cold that we could hardly strike a match. Soon a flourishing bonfire warmed us, however. There in the shelter of the pine trees we cooked our potatoes, bacon, and eggs. Food has seldom tasted better to me.

In the first four grades we read from the Beacon readers and certain other series. Dick and Jane had not yet made their appearance. As I recall the primers of that era, however, I am sure that they were no more exciting than materials currently used for reading instruction which are frequently the object of much criticism. As for readers above the primer level, I would not guarantee that they provided the best instruction. I found them to be interesting and exciting for the most part, however.

To study his reading lesson, a pupil read silently to himself. When he came to a word he did not recognize, he carried his book up to the teacher who might be conducting a third grade arithmetic class or a fourth grade geography class. The pupil would point to the unknown word, and the teacher obligingly pronounced it for him. If a pupil contacted the teacher too frequently for such help, he was reprimanded and told to figure out some of the words for himself. To recite, the pupil read the lesson aloud to the teacher. We read "Red Riding Hood," "Three Billy Goats Gruff," and other stories in the **Beacon First Reader.** As children our initial criticism was that the characters talked, and anyone knows that wolves and goats cannot talk. Our teachers and parents managed to convince us that those stories would be interesting if we gave them a chance, even though such animals do not really talk. Generally speaking, we found this to be true.

In the second reader we read "The Bremertown Musicians" and "Rumpelstiltskin." We never thought to question why a king could be deluded into believing that the poor miller's daughter could spin straw into gold. Why then was her father a poor miller? We read "Jack and the Beanstalk" in the third reader, and in the fourth we read "Bluebeard." And today they talk about violence on television!

We read "Rip Van Winkle," "The Great Stone Face," "Evangeline," and "Enoch Arden." Occasionally we had to memorize a poem, but I fear that too frequently we got little of the meaning which Tennyson, Kipling, or Wordsworth meant to convey. Poetry was often committed to memory to meet a requirement and soon forgotten. Other verses were sometimes memorized for their humor, for patriotic reasons, or for their moral value, for example, to teach us that if we worked hard we would be successful.

This brings to mind some of the arguments we heard during the 1960s regarding the work ethic. To argue that work is a virtue for its own sake is ridiculous; yet to a degree I must defend the work ethic. I have never known anyone to make a railroad tie, write a book, build a house, raise a crop, or perform a valid research experiment without work. On the other hand, I completely reject the notion that by working hard one will always capture the bitch goddess of success. We Linns worked hard for what we acquired, but I have known many tie hacks who worked harder and were more capable than I who ended up with nothing.

The assumption that good poetry was written to express morals was held by many students even after they enrolled in college, much to the exasperation of professors of literature.

In the early grades we were taught that in Holland people wore wooden shoes, the men wore baggy pants and smoked long-stemmed pipes, and that the women wore white caps. We were also told that the Dutch did their traveling in the wintertime on skates. There was, of course, an element of truth in what we were taught; but from the description we received of life in Holland, we would hardly have recognized a scene in Amsterdam as being typically Dutch.

Eskimos were a small people, we were told, about as tall as our twelve and thirteen-year-olds. They ate raw meat, drank blood, lived in igloos made of blocks of ice or snow, and constantly experienced extreme cold. I learned that Alaska was in the far north. Eskimoland and Alaska were one and the same to me as a boy. I envisioned scenes in Alaska in which there were igloos, small Eskimos everywhere, and Santa Claus in their midst. I lost my faith in Santa Claus in the early grades, but until I was about twelve-years-old I clung to the notions about Eskimos and Alaska. I was destined to live in Alaska later for six years. Winters were no rougher there than those I experienced in Wyoming, and the snow was no deeper than it had been on Cottonwood Creek. The temperature never dropped as low as it did in Big Piney. I never saw an igloo.

And, of course, in the lower grades we drew and cut out pictures of cherry trees and hatchets in honor of George Washington.

History was a most interesting subject to me. Patriotism was always emphasized. Textbook authors and teachers did a superb job of rationalizing the Mexican War. When it came to the Civil War, I had a rough time distributing my sympathies. The north was right. Slavery and secession were wrong. Lincoln and Grant were great heroes. My own dear Georgia-born grandfather, however, had some views which I found hard to reconcile with what I was taught about the Civil War.

In the sixth grade I was told by my teacher that I.W.W. stood for "I won't work." It was not until I was a high school senior that I learned that it was the abbreviation for Industrial Workers of the World.

Geography was sometimes interesting. Frequently, though, it consisted largely of memorizing the capitols of the forty-eight states, the names of six or so of the larger municipalities in Wyoming and what each was noted for, and, later, the capitols of the countries of South America and Europe. These were facts which had to be memorized for examinations and all too often were soon forgotten.

What else did we learn in the one room schools? Reading, writing, and arithmetic, of course. A variety of subjects were prescribed by the State Department of Education which also provided a rigid course of study for each subject. The curriculum from the fifth grade through the eighth included reading, arithmetic, language, spelling, history, civics, geography, hygiene, and agriculture. Nature study, drawing, and music were sometimes included; but as they were considered to be of secondary importance, they were taught at the discretion of the teacher.

When I was in the sixth grade in 1924-25, we were required in certain subjects to pass a final examination sent out by the county superintendent's office. Finals were provided by the State Department of Education for all pupils above the fifth grade the following year. The examinations had to be given by a school administrator who was not the pupil's own teacher. Pupils had to go to Big Piney, Daniel, or Pinedale to take the examinations.

That was the last year I went to school in a tie camp. The company moved its operation to Smith's Fork in 1926. Enrolled in the eighth grade in a four room school in Pinedale, Wyoming, I stayed with the editor and publisher of **The Pinedale Roundup** and worked in his shop after school and on Saturdays. I went to high school in Pinedale and later in Mountain View, Wyoming, from which I was graduated in 1931.

I have often felt that I learned much in the small rural schools I attended. Some of it was good, and some of it left me with much to unlearn. This was probably not peculiar to the small schools which I attended. I cannot blame the schools for the prejudices which remained with me for a number of years. I seriously doubt that critics of modern education who speak fondly of the "good old days" would be satisfied with the tie camp schools of the 1920s. I considered myself fortunate just to be able to go to school. Like life in the tie camps generally, school was not easy. It suited me then for the most part because I had nothing with which to compare it. I would not be content with it today, but I have fond memories of certain features of my early schooling.

CHAPTER NINE

Going Places in the Timber

Before moving to the tie camp, we relied on horses for transportation as did most residents of the Piney country. To get to a neighboring community for a dance, a show, or a rodeo, we went by wagon. If the occasion were to keep us away from home overnight, we carried tents, bedrolls, and camping equipment with us. If one person or a couple needed to make a short trip, he or they usually rode horseback. Some of the people owned buggies or small springwagons. A few of the more prosperous had automobiles which could get them wherever they wanted to go during much of the year.

The tie camp was almost a world unto itself. From mid-November until spring a trip to town meant riding for fifteen miles or so on the freight sled and then completing the journey by truck. The road from the camp down to the lowlands during those months was covered by five feet or more of snow. Raging blizzards frequently covered the sled trail with snowdrifts, making traveling slow and rugged. Leaving at daybreak, people traveling by sled reached the ranch which served as a stopover by nightfall. The journey resumed early next morning; and after reaching an area where the snow was sufficiently shallow, the trip would be completed by truck. If the road conditions were fairly favorable, the traveler would reach Big Piney by evening of the second day of the trip. He could not return to the timber until the beginning of the freight run a week later. Few tie hacks could afford spending that much time away from work. Most of them, therefore, figured on remaining in the camp from late fall until after the spring thaw.

From the time of the first heavy snowfall until after the spring thaw there were no wagons or automobiles on the road. Anyone who had a car put it under shelter, propped it up on blocks, drained the radiator, and removed the battery. Any traveling after the first heavy snowfall was done in sleds or on skis. If we went anywhere on one of the go-devils, hay covered with an old blanket or canvas was often spread over the floor of the shallow box which was fitted on sled runners. We could then sit in comfort on the sled.

Sleighriding was fun. Seated on the sled and dressed warmly, we would set out for the commissary or one of the neighboring camps on a calm clear day. For a short distance we rode through the valley in which our camp was located. Nothing was visible in any direction except the snowcovered terrain and the pine trees with their green branches tinged with snow glistening in the sunlight. The stillness was pierced only by the jingling of sleigh bells, the squawking of jays, and the occasional chattering of a pine squirrel. Our trip took us through the forest over a narrow road protected from wind and drifting snow.

Occasionally our destination would take us off a frequently-traveled road. If there had been a recent storm, this might require plowing through drifts. Sometimes the horses would sink into snow up to their bellies. A teamster hauling grub to one of the camps or making an absolutely essential trip faced the toilsome and time-consuming task of shoveling snow from the road with a scoop shovel. The snow on the roads over

(Nils Edin, ca. 1936)

Horses bogged down in soft snow

which heavy loads of ties were hauled, however, was usually firmly packed.

When we were at the Nylander camp, we set out on one occasion to visit a family two miles away. We ran into deep snowdrifts, and Walfred feared for the welfare of the horses. While we children were disappointed not to complete the trip, we did not protest when we turned around and returned home. We could not stand the thought of old Steel or Charlie becoming ill and dying. We made the trip on skis the following week.

Skiing was basically a means of getting somewhere. When a man worked ten hours a day at hard physical labor, he did not feel the need for exercise; and he seldom went sport skiing in his time off work. Occasionally a man would spend a half-hour or so skiing purely for fun. Scandinavians, especially the Norwegians, were highly skilled skiers. In the winter it was not uncommon to see a man sliding down a hill on skis at breakneck speed, carrying his saw under one arm and his broadax over his other shoulder.

Travel was easier in the summer. Unless one had a car, however, a trip as far away as Kemmerer whether for shopping or surgery meant a person had to ride on the mail truck from Big Piney to Name's Hill, have his noon meal at the stopover there, transfer to another mail truck, and go on to Opal. After spending the night there, he took the train to Kemmerer.

Few of the tie hacks owned cars, but many wanted them. Pete, a large husky Swede, saved his money for several years, went down to Lyman, bought a Ford, and gave the dealer a check for payment in full. After taking possession of the car, Pete got hold of some liquor; and he and a few friends celebrated his purchase. They drank for awhile, and then the other fellows clipped the word Chesterfield from a cigarette package and pasted it over the word Ford on the front of Pete's car. Later one of them said to him, "Say, Pete, T'ought you vas gonna buy a Ford."

Pete answered, "Ya, shore. I buy Ford," and looked proudly at his car.

Pointing to the word Chesterfield, the fellow said,

"Vat's that Chesterfield?"

"Vell, by God!" declared Pete. He lost no time in getting back to the garage. "Vere's that sheck?" he demanded of the dealer.

"What's the matter, Pete?" the dealer questioned.

"That ain't no Ford you sell me," Pete said. "It's a shesterfield."

The dealer convinced Pete that the car was indeed a Ford. Pete kept the car and was very happy with it.

When we returned from a trip to Idaho in the summer of 1922, we boys were excited and tickled to learn that Dad was negotiating for a second-hand car. I am unsure as to how he financed the deal. I believe he sold or traded some of his horses which were of no use to him since his freighting days. At any rate, Dad caught a ride down to Marbleton and completed the purchase. The car was a huge old Chalmers. After getting possession and having a very brief bit of instruction in driving and care of the car, he started to drive it home to the tie camp.

Dad left Marbleton with the car in the late afternoon as he was in a hurry to get back to work. The old proprietor of the Marbleton Hotel tried to get Dad to remain for another night and drive home the next morning. He watched Dad drive out of sight, hoping and praying that Dad would get home safely. Knowing that Dad had just learned to drive and was setting out without lights over rough hazardous roads, he said later, "I wouldn't of rode with him for the Marbleton Bank and all that's in it."

Darkness fell as Dad was driving, and he suddenly found himself out in the sagebrush. As he attempted to get back on the road, Charlie Bracket, the Forest Service supervisor, and his wife overtook him. They went ahead in their new Model-T Ford with lights, and Dad followed them. He made it home all right.

It was a big thrill to have a car. When Dad bought the Chalmers there were only four other cars in the tie camp. They were owned by the Forest Service supervisor, the camp manager, one teamster, and one tie hack. We could make frequent trips to the commissary and the other camps and to the south fork of Cottonwood Creek where there was better fishing.

Operating the Chalmers had its problems. The car had a "Hot Shot" battery consisting of four dry cells. This kind of battery did not provide enough power to operate the starter, the horn, or even the lights. To start the motor, it was necessary to connect the battery and crank the car. Dad sent for a storage battery. Installing one in those days was a rather complicated task. We drove to the home of the other tie hack who owned a car. Dad and he spent the entire day installing and connecting the battery so that it would operate the motor. They never did get it to operate the starter, the horn, or the lights. As the man backed the car out of his yard, the axle broke; so he took us home in his car. Dad had the Chalmers towed later by team to the Nylander camp where we were living. It remained stored in the shed until the following spring.

Dad sent for a new axle for the immobilized Chalmers. A man in the camp who had once worked in a Chalmers garage helped Dad install it. Again we were able to travel by car until a spring broke. Fortunately, it was possible to have the spring repaired in Big Piney.

In the fall of 1923 when we were at the Nichols' ranch, Dad traded the Chalmers to a Big Piney mechanic for a second-hand Overland. The Overland was a smaller car, had apparently had good care, and burned less gasoline. Prior to 1923, as I recall, one did not speak of the ignition. On the dashboard of the Overland was a button which one pressed to "turn the batteries on." It was necessary to press that button before the motor could be started. People sometimes carelessly left the batteries turned on while the motor was not running and ran the battery down. Getting a battery recharged was not easy in those days. Usually it meant hitching a team of horses to the car and towing it to get the motor started and then keeping the motor running until the battery was sufficiently charged. If the engine died, the team had to be hitched to the car again. Sometimes the car would be parked at the top of a hill once the motor was started. Then if the engine died, the brake was released and the car coasted until the motor started again.

Our Overland had an ultra-modern feature. If the battery were left turned on and the motor was not started within a minute or two, the mechanism would become heated, sound a buzzer which resembled the horn honking, and the button would pop out so that the battery was no longer turned on. It worked beautifully, except during cold weather when the car had to be cranked. Dad or Mother would press the battery button, get out, and start cranking. About that time there would be a beep, and the battery would automatically shut off. With the mechanism thus heated, almost as soon as the battery was turned on again there would be another beep; and the motor could not be started until the mechanism had cooled down. It presented a real problem on a cold Wyoming day.

We drove the Overland until late fall. The following spring the folks traded it in on a new four-cylinder Buick, probably the last four-cylinder Buick was to make for many years to come. They drove the Buick from 1924 until 1929 when they traded it in on a Model-A Ford.

When getting ready to make an auto trip in the timber, the first step was to fill the radiator. Often it took many trips to the spring because we drained the radiator every night except during the warmest summer months. Many a radiator or block had been cracked because the temperature dropped to freezing. With the radiator filled, a can or pail in the car with which to get water if the radiator overheated, and gasoline in the tank, we were ready to travel. Dad sat behind the wheel with Mother beside him on the front seat. We boys and sometimes our dog sat on the back seat. We always dressed for the weather as the car had no heater.

The windows were never rolled up because there were none. Most of the cars then were not enclosed. Occasionally in a magazine we would see a picture of an ultra-fancy car which was enclosed. One of the adults would murmur that he would not think of buying such a car. In the event of an accident, it was feared, one might get cut by the shattering glass. We never had an enclosed car until we got the 1929 Ford. For our previous cars we had side curtains. They were considered a blamed nuisance, but we would put them up if the weather demanded it. They did keep snow or rain from blowing into the car.

Setting out on an ungraded dirt road, we would drive through the forest. Dad was constantly on the alert for tree stumps. A stump could knock the pan from under the car. Most of the streams did not have bridges. Sometimes the engine would drown because the water was too deep when we forded a creek. More often we would have trouble getting up a hill, especially if the ground were wet from rain or snow. On such occasions or when our car was stuck in a mudhole, Mother would drive while the rest of us got out and pushed. The wheels might spin and splatter us with mud and water. We often kept a shovel in the car during wet weather in case we got stuck. By our digging in front of and behind the rear wheels and tamping some dry brush in the ruts, the wheels usually got traction and we got out of the hole.

Even in dry weather it was often difficult to get up a steep hill. The engine would sometimes die before we got very far up the hill. One of us would get out and place huge rocks behind the rear wheels while Dad held the car in place with the brake. He would then release the brake and try again, hoping the car would not coast back down the hill. Maybe after repeating the process four or five times, we would reach the top of the hill.

To drive more than thirty miles meant leaving the timber. It would be even farther before we reached graded roads. When driving that far one could count on at least one flat tire. There were not many service stations in western Wyoming, and garages were few and far between. We always carried tire patching equipment in the car. When a tire went flat we jacked up the wheel and removed the tire and rim. We pried the tire off the rim with a tire iron, removed the inner casing and inner tube, and searched for the leak. When we located the leak, we patched the tire, put a little air in the tube with a tire pump, and made sure it was not leaking. We replaced the tube and put the tire on the rim. After replacing the wheel came the fatiguing task of pumping with the hand pump until the tire had sufficient pressure. Letting the wheel down on the ground again, we put all the equipment back in the car and drove on, hoping not to have another flat for awhile. Although we always carried a spare tire, we usually repaired a flat on the spot because it had to be done eventually.

We never considered traveling by automobile to be tiring or uncomfortable. It certainly beat riding in a wagon, both in terms of comfort and the distance one could cover in a day. I remember riding in the Buick with the folks and others to Rock Springs in the summer of 1925. We had beautiful weather, and it was a great day for traveling. As we went through Pinedale, Dad noticed from a sign that it was 102 miles to Rock Springs. He said jokingly, "Gee whiz! That'll take two hours." He really got the horse laugh. Unless a man wanted to commit suicide, he would not drive that fast. As it was, we cruised between thirty and thirty-five miles an hour, stopped to eat some sandwiches after driving about fifty miles, and arrived at Rock Springs in the early afternoon. We had only one flat tire on the way.

CHAPTER TEN

Going on the Drive

I first witnessed a tie drive in the spring of 1920. For a couple of weeks the drive had been the subject of conversation in the Moline camp. Yawn, the old tall, lanky Swede, was going on the drive, but Nils, a husky, little bald Dane, was not going on the drive. He liked to make ties at that time of the year. Carl, old and slightly feeble, was going. Young Andrew was going on his first drive as the cook's flunky.

Soon after the drive started, we children scampered down to the landing where thousands of ties had been unloaded on the bank during the winter. It was exciting to see the water flow down the creek which had been so recently frozen and snow-covered. The eight feet of snow had been especially slow in melting that spring. Huge drifts remained on the creek banks and against the tie piles. A few men were shoveling it out of the way with scoop shovels. Others were standing on the piles, dragging ties with pike-poles and pickaroons, and throwing the ties into the water to set them afloat. Still others were working farther down the creek, pulling ties away from the banks to keep them moving.

(Nils Edin, 1913)

Rear gang with pike-poles ready for work on Mill Creek

These ties would float down Cottonwood Creek to where it joined Green River. They would be stopped there by a tie boom built of logs at the mouth of the creek. In the spring the water level of Green River was too high for driving. Ties floated down the river then might drift out into fields so that when the water finally lowered they would be far from shore. The ties would be held at the boom until July when the drive would resume.

From there the ties would be driven to the town of Green River, Wyoming, where a large boom stopped them. They would then be loaded onto freight cars to be transported to Laramie to the pickling plant. Here they would be dipped in creosote to preserve them and then shipped to wherever they were needed along the Union Pacific Railroad.

Drivers were awakened at six o'clock by the cook's call "Roll out!" A driver would hastily pull on his woolen shirt, heavy trousers, woolen socks, and hobnailed calked boots. He would warm a half-gallon can of water over a campfire and wash his face. On days that the camp was to be moved he would take down his tent and roll his bed.

The cook on the 1931 Smith's Fork drive adamantly refused to make hot cakes over a campfire for forty hungry "river hogs." For breakfast he served fried potatoes, bacon, boiled eggs which were often cold by the time they were cracked, cereal, canned fruit, and cold baker's bread. When a man tried to butter that bread on an early spring morning in the Wyoming mountains, it often brought forth a few Swedish cusswords. On other drives cooks like Ernest Nelson and Big Gus served hot cakes or French toast almost every morning.

At seven o'clock the drive boss called "All out!" and the ten hour workday was underway. The first step in a drive was breaking up the landings, that is, getting all the ties from the piles into the water. From then on the task was to keep them floating. Because of the freezing temperatures some mornings, there was not sufficient water in the creek for ties to float. A half--day's work might be lost while men lounged around camp waiting for the creek to rise sufficiently so they could go back to work.

After pulling ties for five hours, the men heard the call "All off!" Drivers would go to the camp for the noon meal and a rest until one o'clock. Fellows would file past the Dutch ovens and stock pots, filling their plates with meat, potatoes, canned vegetables, and pudding. After another five hours of work, a man was usually pretty tired and hungry when the "All off!" call came at six o'clock.

Very often supper included split pea soup which really hit the spot for a worker after he had waded through cold water for hours. This soup was made by dumping a quantity of dried split peas into a stock pot, adding chunks of salt pork and water, and boiling the mixture over the campfire. Meat, potatoes, more canned vegetables, and rice or tapioca pudding would be served to the men, and there was always plenty of potent boiled coffee available with every meal.

There were a few outsiders who would show up at a drive camp at mealtime and take on a free meal. They included hoboes who were beating their way across country during those hard times. If the drive were camped near a small community, young people from town would show up, get in line, and help themselves to food. The boss and the cook never protested so long as they got at the end of the line to make sure that the drivers got their share first.

When camped on ranch property, people from the ranch would occasionally show up and enjoy a driver's meal. They were usually made to feel welcome. The bosses were pretty tolerant of ranch people who often resented having the drive go through their land and the damage done to their irrigation ditches and headgates. Upon occasion some people abused the privilege. One ranch woman on whose property we camped for several days ate all her meals at the drive camp. Every morning she put away a quantity of hot cakes which she ate with her knife.

Perhaps the worst offender was an old moocher who lived near Daniel. The drive was camped for several

(Nils Edin, ca. 1927)

A camp on the Smith's Fork Drive

days at the boom where Horse Creek meets Green River. Mr. Kringley (not his real name) walked some distance from his home at mealtime each day to eat a hearty meal at the drive camp. After they broke camp and most of the drivers were gone, several men stayed at a ranchhouse while they did some work on the boom. Mr. Kringley continued to join them at mealtime. In exasperation the men finally agreed that the next time he arrived no one would invite him to sit at the table and join them. Kringley sat and watched the men eat. No one offered him any food. There was a good roast on the edge of the stove near the table. At last the temptation became too great for him. Mr. Kringley picked up a slice of bread from the table, took a slice of meat from the roaster, and made himself a sandwich. With that, one fellow's heart softened; and he poured Kringley a cup of coffee.

John Wasburg, one of the men, was in Daniel a few days later. It was nearly noon, and he started walking toward the ranch where the boom was located. Kringley met him and asked where he was going. John said it was about time to go to dinner. Kringley said, "Say, John, come on over to my place and eat with me today."

That was indeed a switch, and John promptly took Kringley up on his invitation. As they walked toward Kringley's house they came to the general store. Kringley said, "Say, John, you got any money."

John replied, "Ya, I got a little. Vy?"

Kringley asked, "You got a dollar?"

John said, "Ya, I guess I got that much."

Kringley said, "I'll tell you! Let's go in here, and you buy some bread and lard. I'll buy some eggs, and we'll go over to my place and have a real feed."

Later that afternoon one of the men came into the ranchhouse to discover John raiding the cupboard with both hands as he gorged himself. "What's the matter, John?" he questioned. "Didn't you have any dinner?"

"Ya, shore," replied John. "Had dinner with Mr. Kringley. Had bread and eggs. Cost me eighty cents. Cost him thirty-five cents."

Many situations and events were pathetic as well as humorous; but unless one saw the comical side of life in the tie camps, existence there would have been dull if not tragic.

With the drive underway, the boss would start cutting down the size of the crew although it was necessary to retain a sufficient number to keep the ties moving. On any drive a few men were always sent down the creek on the lead. Their job was to put cribbing at the inlets of slews which branched off from the creek, thereby preventing ties from floating into the slews. As the drive progressed, some ties would drift to one side of the stream or the other. When the water level subsided, they would be left stranded on sand bars possibly twenty or thirty feet from the creek. This necessitated hooking onto each tie with a pike-pole, dragging the tie to the creek, and wading with it out into the current. The process of getting ties back into the current was called pulling rear. When many ties were stranded far from shore or when wet sand made pulling rear difficult, the men would sometimes lay down their pike--poles and start packing.

This meant shouldering each tie and carrying it out into the creek. A few old lumberjacks considered this the only effective way of moving ties any distance. It was said that when Martin Johnson parked his ties in the timber he never dragged a tie with a pickaroon but always shouldered it and carried it to the pile. Inexperienced drivers found packing very rough and knew they could expect sore shoulders and muscles the following day, but they continued to heave heavy damp ties because the boss had an eye on them.

Sometimes it was necessary to pull jam. If the water were shallow over a stretch of creek, the ties got blocked and would jam up to a quarter of a mile. The boss would pull a number of men off the rear gang and get them to pull jam. On the smaller streams a charge of dynamite would sometimes be placed under a jutting bank. This would eliminate the obstruction and loosen the jam.

After the landings were broken up, camp was moved

every day. On the Smith's Fork drive, bedrolls, cooking utensils, and supplies were loaded onto a truck and hauled to the next campsite. Camp was moved by boats on the Green River drive. One boat hauled the cook's equipment and supplies and another boat carried the bedrolls. There was also a rear boat by which drivers crossed the river when necessary to go to the other side.

Clear weather, a high water level, no jams, and light rear kept the boss happy and made work easier for the drivers. The boss always wanted to get the drive over in a minimun amount of time.

As the drive got into the lowlands the weather was warmer. After ten hours of pulling rear, packing, and wading, the men enjoyed sitting around a campfire after supper spinning yarns or perhaps playing pinochle. If they happened to be camped near a town, they might board the truck which moved camp or hauled grub and supplies and go into town to a movie or for a drink or two in one of the bootleg joints.

Every year Ephraim, a Finn, went all the way on the drive. With cash in his pocket, he would go down to Diamondville, Wyoming, visit with some old friends, and have a fling in a bar before returning to the timber. He was a short, wiry little fellow with a big blonde moustache who remained in remarkable physical condition despite his advanced age. After a few drinks in Diamondville, he would stand in the middle of the floor, jump into the air, lift his feet and legs, and land flat on his rear-end.

At the beginning of the drive one year, company men from the Evanston office suggested that Ephraim was too old to go on another drive. Standing on the commissary porch, Ephraim turned a handspring and said to the company men, "Now, son o' mo bitch, do that!" Ephraim went on that drive and subsequent drives.

When I first knew him Ephraim was about eighty years old. He lived alone in a cabin isolated from the rest and managed to make enough ties to feed himself and provide a quantity of meat for his dog Penny. His knowledge of English was quite limited. He sang in Finnish to Penny, and the dog howled. They had quite a duet.

Standing in the creek near the shore on the 1933 drive, Penny splashed water on one of the men working nearby. The man whacked Penny lightly with his pike--pole, and Penny let out a yelp. Old Ephraim cried, "Don't hit that dog! Son o' mo bitch." A long string of Finnish then poured from him.

The man said, "Well, keep your dog home then."

Ephraim responded, "Son o' mo bitch!" More Finnish followed.

Unwilling to knuckle under to the old Finn, the man declared, "I'll kill the son of a bitch if he splashes water on me."

Ephraim repeated, "Son o' mo bitch!" with additional Finnish words. Old Ephraim got the last word.

I went on my first drive in the spring of 1931 just after graduating from high school. It was the Smith's Fork drive which ended at Granger, Wyoming. Old hands received $4.50 a day, and I was paid four dollars a day and board which seemed tremendous to me. As a driver had his bedroll with him, his lodging cost nothing. Times were hard and work was scarce. Just about every tie camp employee and many of the ranchers and others from the valley below worked on that drive.

I bunked with Dad. In prior years he had worked on the drive until it got underway. When the company was not so desperate for drivers, he usually returned to making ties at which he could make more money at that time of year. In 1931 business was poor, however, and the company limited the number of ties a man was allowed to cut. Dad and I both worked until the drive ended about three weeks later. My brother Bill and our young friend Mike McGeath worked as cook's flunkies.

Our first drive camp was at the Utah-Wyoming state line. We ate our meals in the open and slept in tents as we did on the remainder of the drive. Some of the old lumberjacks told me that on log drives elsewhere a good driver always kept his feet dry. He could balance himself on heavy floating logs. A railroad tie was not

(George Linn, 1920)

Beginning of the drive on North Cottonwood

heavy enough to hold a man up and working on a tie drive required that a man get wet. One of the unique features of a Wyoming tie drive was that a man was not considered to be a true driver until, as the men said, he "got his balls wet."

A driver had to wade through water which was very cold at that time of year. When one came in from work wearing wet clothing on a cold evening a campfire felt mighty good. At the beginning I wore a shirt and shorts for underwear which had become the vogue about a year earlier. We teenagers detested long underwear. Seasoned drivers warned me that I would be sorry, but it seemed to me that it was better to wear shorts than to sleep in long-legged underwear which would be wet from wading through water. Old Frank Murray said, "Yes, but wool underwear is warm even when it's wet." I soon learned that those old fellows were right. I was warm enough during most of the day, but during the early morning and late evening I was cold! Worse yet, I nearly froze during my first night in the tent. I soon equipped myself with long woolen underwear which proved to be much more comfortable for sleeping, even though the legs were soaked.

A supply of snoose was always on hand. I had tried it a few times. A fellow passed around a box of Copenhagen. Feeling a bit of a smart aleck that morning, I put a pinch behind my lip and went to work. As I started tugging at ties, I began to feel woozy. I spat out the snuff but still did not feel well. In an effort to get rid of the foul taste I put a stick of gum into my mouth. This made matters worse, so I spat out the gum. A few minutes later I fed my breakfast to the fish. Several fellows upstream were watching me and I fully expected to get razzed, but no one said a word.

The Irishman, as he was nicknamed, and I were walking downstream together a week or so later. He said, "Give me a chew of your snoose."

I did not believe that he used it at all, and I replied, "I don't use it."

"Oh," he said, "You don't use it."

I am sure he knew what had made me sick, and I am

also sure that I have never again tried snoose.

I enjoyed the 1931 drive as I did later drives. The work was easier than making and parking ties. After a year's work in school, it was as good a vacation as hunting, fishing, or camping. Not only that, it paid money which was scarce in those days. Some of the older men did not share my contentment. As I worked with the Irishman, he said, "Well, lad, how much of this kind of life do you want to lead?"

I said, "Oh, I don't know. It suits me pretty well."

"No, my boy!" he said. "Break away from it as soon as you can."

I agreed that I probably would not be content to do this kind of work for the rest of my life, but I told him that it was kind of fun for a change and was a paying job to boot.

"No," he said. "It isn't any fun! You work hard all day. It's dirty! You get yourself all wet. When you come in at night, you can't take a bath and put on clean dry clothes. It's terrible! And besides, look at what you've got to contend with. Dirty double-crossin' . . .!"

I do not know who the Irishman alluded to as "dirty double-crossin'!" As time went on, though, I could understand his bitterness. Many of those old lumberjacks worked for years at the hardest of physical labor. They lived under primitive conditions with almost no luxuries or comforts, saved what money they could, and all they could look forward to was more hard work if indeed they could remain employed during difficult economic times.

The drive moved on to Granger where we were all paid off. There were the usual baits for getting a share of the payroll: a bar serving home-brew beer and bootleg whiskey, a stud poker dealer, and a shack which housed a couple of prostitutes. Drinking, hilarious laughter, and Swedish folk songs could be heard all night long.

Some of the men remained in Granger and worked at loading ties onto freight cars. With a little money in their pockets, the rest purchased a few articles of clothing which were unavailable at the commissary or which cost less than commissary prices. The men then boarded a truck which took them back to the timber to make ties for the 1932 drive.

CHAPTER ELEVEN

The Noble Experiment

Progress during the later 1920s made life a little easier for tie hacks until the economic crisis of 1929. They enjoyed more conveniences and were less isolated during at least part of the year. Even under improved conditions, however, they could look forward to little more than the same old grind of hard backbreaking work. After working diligently month after month, it is small wonder that certain of them drew their wages and drowned their woes in liquor.

The Eighteenth Amendment went into effect during our first year in the tie camp. It did not take long for tie hacks to use up their reserve of legally-made liquor. Booze was sold illegally in the western cow towns. Generally, a pool hall was a bootleg joint. A man could step up to the bar and pay twenty-five cents for a shot. The bartender would fill a small glass with moonshine and draw or pour a glass of water to be used as a chaser. Standing at the bar with one foot on a brass rail, the customer would down his drink and a gulp of water. The bartender would then rinse the glass in cold water and replace it on the shelf. If a group were on its way to the tie camp, it was customary to buy a quart for the road.

During the first year of operation on Cottonwood Creek, trips to town were few and far between; and it was not easy for the men to get anything to drink. Probably no one had a still, and most of the tie hacks would not have known how to use one. Malt syrup and hops were not available for making beer. There were no grapes available for making wine. The problem was solved, however. Men bought packages of raisins at the commissary, dumped them into a keg or crock with some water, and let them ferment in a warm spot behind the stove for a couple of weeks. They strained the mixture through a clean linen cloth or bag, squeezing all of the fluid out of the raisins. This resulted in some pretty potent wine.

Eventually more men acquired stills, and bootleg whiskey and home-brew beer were often available. With a piece of copper tubing some fellows converted pressure cookers into stills. Corn was usually unavailable, so mash was made from corn meal purchased from the commissary. After his work was done in the afternoon, an old Norwegian cook would ski through the forest carrying his still in a gunny sack and stopping at one cabin or another to run off a batch of white mule. He also made liquor probably resembling Scandinavian aquavit from potatoes.

During our third year in the tie camp a fellow living in the Nylander camp set up a still, made whiskey, and sold it for seven dollars a pint. Some of his neighbors and customers heard a rumor that prohibition officers were going to pay him a visit, so one time when he was very drunk they hid his still and all of his products in the forest. After he had sobered up he could not remember where they were hidden nor would anyone tell him.

The prohibition agents did visit the Nylander camp. They drove in pulling a trailer on which was mounted a large cage with barred windows. The bootlegger was at work out in the timber, and his house was locked up. The officers looked around the camp for awhile but drove away emptyhanded. How eager they really were to make an arrest nobody knew.

When liquor was available some men would spare no effort to get a share. Andrew liked his liquor and would occasionally overindulge when work was slack. Sometime he became quite ill afterward. At the beginning of the drive one year he showed up in drive camp completely stoned. "I never vant to take another drink ever," he said. Lying down on his bed, he continued, "I'm a sick man. I t'ing I gonna die."

Shortly afterward a fellow came to Andrew's tent and said, "Hey, Andrew! Someone just brought a jug into camp. Better come and have a shot."

Andrew responded, "Vell, bring it here! I can't come and get it. Can't you see I'm a sick man?"

If there were whiskey in the commissary camp and a man took a bottle to Ole's camp, the tie hacks would drink it and then go to the commissary to get more.

In the winter of 1922-23 a man brought a supply of whiskey into the Nylander camp where we were living. Some of the men took a bottle or two up to old John's camp two miles north. After finishing the bottle off, the fellows left for the Nylander camp to get more liquor. They left old John all alone. The snow was deep and soft, and it was almost impossible to make the trip without skis or snowshoes. They took all that were in the camp.

In the middle of the afternoon, Bill and I were out in front of our house when we saw old John coming. We did not recognize him at first. He had shaved off his big moustache, nicking his skin several times while shaving. He had used two gunny sacks of hay tied to his huge feet as webs in order to get to the Nylander camp. "Hello, Yords! Hello, Bill!" he called. "Know me? I am Yawn!" Singing and dancing over the soft snow with those sacks on his feet, he went on to the house where liquor was available. The following day he had sobered up and wanted to go home but had to look all over camp to find a pair of skis. When sober, he refused to put gunny sacks on his feet.

During our second year in the Bloom camp, the commissary got in a supply of a patent medicine called Tanlac. I do not know what ailment this medicine was supposed to cure. The fellows in one bunkhouse discovered that while it had a foul taste it had a high alcoholic content and bought a supply of it. When a man came in from the cold he would have a nip from his Tanlac bottle for as long as it lasted. He did not seem to get exceedingly drunk from it.

There are many tales about damage done by the tie hacks when they were drinking. Two characters in the Smith's Fork tie camp were Big Vic and Little Vic (not their real names). Although he seldom had any trouble with others, Big Vic was usually recognized as the toughest man in camp except for Allen Westerberg. It was said that at one time in a Salt Lake City poolhall three fellows ganged up on Big Vic who picked one of them up by the ankles and used him as a club to beat off the others. Several men were hospitalized as a result of the encounter.

Little Vic was a short fellow who was usually drinking except when he was working. Those who knew him said that if he had five dollars in his pocket he was happy; if he did not have five dollars, he was happy anyway. He got to partying one time in Mountain View and became disorderly. The local blacksmith, Jesse Clark, was a deputy sheriff. Jesse finally collared Little Vic and locked him up in the blacksmith shop. The latter crawled through the chimney and escaped. Jesse caught up with him later, and Little Vic was kidded afterwards about having scabbed on Santa Claus.

At the end of the drive in Granger, Nils, the old drive boss, and Little Vic were in a bar together. Nils chewed out Vic. Endeavoring to pacify Nils, Vic said, "No, Nils. Here, Nils! Ve have beer." He ordered a couple of beers.

Nils said, "Ya, Vic. All right, Vic!" As they stood together at the bar, Nils lifted his glass, saying, "Skal, Vic!" After drinking about half of his beer, Nils then looked at Vic and said, "But you're a Laplander yust the same!"

Brawls did occur occasionally, and there are authentic stories of scraps during which fellows threw beer bottles through windows. Men might go into town and get arrested for being drunk and disorderly. Rarely, though, did I ever know of them to carry their celebrating to the point of serious destruction of property. Supposedly, a bunch of tie hacks once went into town and tore up the wooden sidewalks with their peaveys. I doubt it. A peavey is a very heavy tool, and a man did not carry it along when he went away for a fling.

It is true that occasionally during a drinking spree some men would become violent. Often a quarrel or fight occurred when one man accused another of stealing or taking more than his share of liquor. During 1922-23 Hugo and Andy shared a cabin in the Nylander camp. They frequently got into scraps. One night Hugo knocked Andy, who was only partly dressed, against a hot heating stove. Poor old Andy's back was so severely burned that he was unable to work for some time.

A terrific battle took place in the same cabin the following summer. Dad went down to investigate the commotion and racket. He discovered that some tie hacks had thrown Andy headfirst into a little cellar under the cabin's floor, had stuffed hay from his bunk around him, and were threatening to touch a match to the hay. No one will ever know whether they would actually have done so. Dad pulled Andy out of the cellar and kept watch until someone else relieved him.

When Dad came home he said that the altercation had begun because the men believed Andy had stolen a jug of whiskey. I recalled that earlier that afternoon an old fellow named Oscar had hidden a gallon jug between our house and our rabbit cage, asking us children not to say anything about it. Old Oscar was a heavy drinker who used to walk around talking to others or to himself in a whisper. We children had not given much thought to Oscar's hiding the liquor as it was a common practice. I told Dad about it, he returned the jug to Andy's cabin, and the drinkers settled down.

Old Andy was an unusual character. He did not drink frequently but would overindulge from time to time. On such occasions he called himself "Drunken Andy." He had some musical talent and often entertained others on various musical instruments, especially his zither. He was not very big but was sturdy and a good tie hack.

While he was not an accomplished cabinet maker, he had ability as a builder and constructed good log houses. Some people living on a ranch once asked him if they might engage him to build a log house on their place. A slightly inebriated Andy replied, "Ya, shore! I build you a house. And ven I build it I don't use any chinking or daubing. I stuff excelsior betveen the cracks. And ven I build it I build it yust like a t'ermos bottle. You varm it up vunce, and it stay varm all vinter."

Andy was kind, warmhearted, unselfish, and would always go the extra mile for anyone. In the summer of 1923 my brother Herschel, who was not quite six years old, was struck on the head by a splinter from a falling tree. When word of the accident reached the bunkhouse at the Nylander camp, old Andy ran two miles to the commissary to have the Big Piney doctor telephoned. By the time Herschel had been given first aid, loaded into a car, and taken to the commissary, Andy, quite out of breath, was there and had delivered the message.

Oscar seemed to function as a worker when he was not drinking, but if there were any liquor on hand he drank as long as it lasted. Some of his fellow workers said it ought to be a penitentiary offense to give Oscar a drink.

When my folks were living at the Smith's Fork commissary camp and my brother Jimmy was about three years old, Oscar, who had been drinking, gave Jimmy a five dollar bill. Mother promptly retrieved it and returned it to Oscar, telling him that he worked too hard for his money for him to give it away. She added that he might give Jimmy a nickel or a dime sometime but that he should not give him five dollars. At about one o'clock that morning Dad and Mother were awakened by a loud knock at the front door of the cookhouse. Dad turned on the flashlight and went to the door. Old Oscar handed Dad a candy box, saying "This for the kid." When Dad opened it he found a battered one cent piece carefully packed in a quantity of tissue paper. Some years later Oscar left the timber, and I heard that he was working as a swamper in a barroom.

When we were going to school at the Burns camp, we children sometimes played in the large hayshed adjoining the barn. Occasionally old Nils would show up in camp with a jug of whiskey and invite a few others into the hayshed for a drink. On these occasions he would tell us children to leave. "Kids! Get out of here. Got no business here." I do not know whether it was because he felt that his drinking might set a bad example for us. While he was talking with another fellow, he said, "Kids! Get out of here. You see, now ve're talking somet'ing ve don't vant you to hear."

A number of the men kept a little liquor on hand and after work would have a drink or two but seldom overindulge. If a man had to make a trip a great distance on skis, he often carried a pint bottle with him from which he would take an occasional nip. Others would abstain and work hard for many months. If moonshine were available, they might go on a real spree lasting for a week or two. This was disturbing to the company management because it wanted the men to spend their time making ties.

Like a number of employers elsewhere, the bosses sometimes cussed out the bootleggers and threatened to have them closed. Occasionally a place of business in some of the small towns down the valley would get raided. Often the proprietor would be tipped off in advance by someone who had observed that "the law" was in town. Officers might search a place, seize some liquor, and arrest the proprietor on several counts. When the case came before the court, the bootlegger would plead guilty to one count in return for having the other counts dropped. After paying his fine, he would continue to operate. A few feared conviction on a second offense and sold their businesses to others. In any case, liquor was usually available.

According to certain of the old topers, the hardest part of a drunk was sobering up afterwards. If a man woke up with a hangover, the usual remedy was to take a drink and then a second drink to hold the first down. If his liquor supply was exhausted, he might resort to vanilla extract or, in the later years of prohibition, to a peppermint flavoring with a high alcoholic content available at the commissary.

Prohibition seems to have been about as effective in the timber as elsewhere. I do not recall ever hearing any of the Swedes discuss prohibition as a political issue, but it was discussed quite frequently by others. Very often the person defending prohibition would preface his argument with the words "Well, now, while I do take a drink once in awhile myself . . . ." Perhaps, as someone suggested, he staggered to the polls and voted dry.

Repeal of prohibition brought some changes in the tie camps. It is doubtful if the men drank more after repeal. They bought their liquor from licensed dealers and brought it up to the timber. There were fewer fights over whiskey. The heavy drinkers continued to overindulge and miss work for prolonged periods. With liquor readily available, a hardened drinker did not feel that he had to take advantage of a supply which was hard to come by. In any case, I am sure the quality of the liquor was superior and drinking it was less hazardous than imbibing the moonshine which had been available during the so-called dry years.

CHAPTER TWELVE

Three Graves

A visitor entering the Wasatch National Forest south of Robertson, Wyoming, and proceeding across the state line may notice three graves near the road on the Utah side. The first grave is that of Ole Olson. An aged man when I met him in 1922, he claimed then to be eighty-two years old. Some said that he had given the same figure as his age for several years and continued to do so for years thereafter.

One frequently hears of tow-headed Swedes who are generally thought to have fair complexions. Old Ole was quite the opposite. What was left of his once dark hair was turning gray, and he had a long, full black beard.

I remember conversations between Ole and Chris Conchef who worked with Dad during the summer of 1922. Chris once told Ole that he intended to buy a car. Ole said he was crazy. "That ain't nothin' for a tie hack to have," he said. "You buy a car and then you got to buy oil and gas and tires and grease and all kinds of yunk. If you gonna buy a car," he continued, "you ought to get vun that got vings on it and go flyin' vay up in the air. I saw vun of them vunce, and it looked yust like a little hawk vay up in the air." He added, "But I don't even vant vun of them kind. Yust an old pair of overalls and a can of pork and beans! That's all I need."

That did indeed seem to be all that old Ole needed. He lived alone in his cabin. At his age he could not make many ties in a day, but he managed to get by. He was fond of talking of his youth. "I've had a long time of life," he would say. "I might go in a minute. I drink great deal of viskey in my young days, but I don't like to get out much no more. I yust like to stay right here by myself. Vat I got right here is good enough for me."

He once said to me, "I vent to school ven I was a kid, and I tell you I vas ornery. I done all sorts of bad things. And the teacher, I tell you, she couldn't handle me either." He continued, "But don't you do that vay ven you go to school. That ain't no vay to do. I tell you, vait until recess. Then you can be ornery and yust raise Cain. Don't do that vay in school, though."

Ole had a good .22-caliber rifle, and his visual acuity was still sharp enough that occasionally he would knock off a rabbit for food. He also let it be known that the gun was his protection in case anyone tried to bother him.

Chris borrowed Ole's gun and shot a few rabbits. Ole allowed him to use the rifle after giving him careful directions regarding its care and warning him not to scratch it up. Chris was absent from his cabin later and left the gun there. Upon learning that Chris was not at home, Ole panicked as he feared that Chris had moved away and taken the precious gun with him. Few people in the timber locked their cabins in those days, and Ole found the rifle in Chris's unlocked cabin. When Chris came home, Ole gave him a stern lecture. He was afraid his gun might be stolen if left again in an unlocked cabin. From then on, Ole would lend Chris his rifle only if the latter left twenty dollars as security.

Ole wore overalls and a jumper in the summertime. During the cold season, he wore heavier trousers, a heavy woolen shirt, and a warm jacket. He wore no underwear, and people were amazed that he did not freeze. Often his clothes became pretty ragged, and he mended them himself until he broke his only needle. He was heartbroken over that because he had used the needle for many years. When telling others about it, he added with a tear in his eye, "It was all steel, too."

About 1928 Ole's health failed, and it was obvious that he could no longer work. The company bosses offered to find him a place in Evanston where he could live and have care. Ole refused. He said that if they were tired of having him around he would shoot himself. His only desire was to continue to live in his own cabin.

"W" hauled a load of grub down to Ole's cabin near the state line. He spent a whole day sawing and splitting firewood for him and brought such clothing as Ole would need, including two suits of underwear, from the commissary. Old Ole spurned the underwear. "That stuff's for vomen and kids," he said. He refused to wear it.

Ole improvised three different devices for locking his door from the inside. He still had his gun and made it clear that he wanted no one trying to get in or fooling around. His greatest fear was that someone would try to move him to Evanston.

In the spring of 1928 tie hacks living nearby noticed that no smoke was coming from Ole's stovepipe. They were reluctant to go to his cabin, fearing that he might take a shot at them. They sent word to the commissary, and "W" went down to investigate. He rapped at Ole's door and got out of the way in case the old man should shoot through the door. Getting no response, he broke the locks on the door and found Ole's dead body. Ole was buried at the state line.

Two weeks later when Dad finished work one evening he recalled that for a few days he had heard no sign of activity on the strip of timber worked by old Jack Rose nor had Jack been to the commissary to pick up his mail. As Dad passed through the timber where Jack had been working, he feared the worst. All of Jack's ties had turned yellow and were obviously not freshly--cut. At Jack's cabin Dad could see swarms of flies through the windows. When he opened the cabin door the stench impelled him to turn and leave as promptly as possible. He was reluctant, however, to report a death unless he was sure. Going over to Jack's bunk he beheld the gruesome sight of Jack's partly decomposed body.

Dad reported Jack's death at the camp manager's office. I do not know what provision Utah law made at that time regarding coroner's inquests, but the county sheriff decided to investigate. Perhaps he was suspicious because two deaths had occurred within such a short time. When the sheriff arrived at the cabin with the mortician and the company's general manager, he declined to go in.

There were indications that Jack had intended to be at his cabin for some time. A quantity of food had been prepared and a batch of doughnuts had been spread on the table to drain and cool. A revolver was found on the bunk with Jack's body, and it was assumed that he had shot himself through the head. What motivated Jack to take his own life remains a mystery. He was buried at the state line next to Ole.

Between the state line and the site of the old commissary camp, there is a small clearing in the forest which is labeled "Suicide Park" on Forest Service maps. The body of Charlie Matson, a Finn, was found there in the winter of 1930-31. Charlie, who lived in a nearby cabin, was at times a heavy drinker. He had been on a drunk and had run out of liquor. Noting his desperation and fearing that he might do something drastic, some of the fellows took his guns out of his cabin and hid them. Unfortunately, that did not stop Charlie. He went out into the forest and slashed his throat with a razor. The spot where his body was found by men hauling freight to the commissary has been called "Suicide Park" ever since. He, too, was buried at the state line.

A short time after Charlie's death a copy of the **Daily Worker**, a Communist publication, found its way into camp. It carried a little item stating that Charlie Matson had committed suicide after his wages had been cut by the Standard Timber Company without a corresponding cut in commissary prices. Men in the camps who might have been sympathetic to this paper's editorial policy knew that this assertion was ridiculous. The cut in wages occurred after Charlie's death. I honestly believe that the impression of the **Daily Worker** I thus received was a major factor in preventing me from every considering membership in the Communist Party. It has also caused me to disagree with those who would censor subversive publications as the Matson article did exactly the opposite of what was intended.

The names of Ole Olson, Jack Rose, and Charlie Matson on their grave markers are nearly obliterated with age; and there is nothing to show when, where, or how these men lived and under what circumstances they died.

CHAPTER THIRTEEN

The Lean Thirties

The last complete year that I spent in a tie camp was that of 1925-26. At the end of that summer the Standard Timber Company moved its operation to Smith's Fork. Our house on the south fork was sold to a rancher who tore it down, hauled the logs to his ranch, and reassembled the house. Any buildings not disposed of in this manner were burned by the Forest Service after the first snowfall when there was no danger of forest fires. Today little evidence remains indicating that a tie camp ever existed on Cottonwood Creek.

After the company's operation moved to Smith's Fork, tie hacks and their families enjoyed certain advantages which they had not had on Cottonwood Creek. The commissary camp was located in Utah, twenty miles south of Robertson, Wyoming, and just three miles from the state line. Robertson consisted of little more than a general store and post office but was only about thirty miles from the railroad. Wages were comparable to those on Cottonwood; and as the camp was nearer the railroad, prices were a little lower. People who had cars usually put them in storage for the winter; but during certain years, depending on the depth of the snow, they could drive between the tie camp and town during much of the winter. Thus during much of the year people could get to the towns in the Bridger Valley with comparative ease.

There were more families in the outlying camps, and they got together at the commissary cookhouse for Saturday night dances and occasional parties. In 1933 the residents of Smith's Fork put up a log building with a dance floor made of planed lumber. Parties and dances were held there until the company ceased operation on Smith's Fork in the late 1930s. Music was usually provided by accordion players Sture Olson or John Pearson. Occasionally an orchestra from down in the valley was engaged, and until three o'clock in the morning people would dance the waltz, the fox-trot, the hambo, and the schottische.

Except for a few old tie hacks who lived alone in their own cabins isolated from others, life was far different from that of a decade earlier. Winters were not so long and evenings were much less lonely than in the Bloom camp of 1920-21.

Times were rough economically for tie hacks in the 1930s. People who for years had produced railroad ties reaped little benefit from their toil, and now they could look forward to nothing but unemployment and poverty. When I was a senior in Mountain View High School, I was stunned when I heard just before Thanksgiving in 1930 that the tie camp had closed down, that the commissary porch was covered with tools, and that men were leaving in droves.

For more than a year people had talked of depression, hard times, and unemployment; but such talk seemed foreign to me. We had felt pretty secure. Bill was a junior in high school, and Herschel was a freshman. It cost us seventy-five dollars a month for board and room; but Mother was running the cookhouse, Dad was making ties, and we all earned money during the summers. Conditions had been better for us than in the Moline and Bloom camps a decade earlier. We boys did

(George Linn, 1932)

Horse Creek Tie Camp

First row: Nils Edin. Second row: (left to right) Jack Morgan, Windsor Hackler, Nellie Hackler, unidentified woman, and Victor Hackler.

not need to ask for summer jobs. We just got our tools and went to work. I could not imagine Dad being out of a job.

We went home for the Thanksgiving recess and learned that the outlook was indeed grim but not quite as bad as we had feared. Most of the employees had been laid off, but a few would remain. The cookhouse would continue to operate. Tie hacks who had been with the company for many years would be retained at least for awhile, but the number of ties they could cut would be limited.

Apparently the company had nearly enough ties to fulfill its contract with the Union Pacific Railroad for 1932 and 1933, and there might be no contract for 1934. Dad and Mother told us that we would have to scrimp and to spend no money for anything that was not absolutely essential.

At the end of the school year, Dad, Bill, and I went on the drive. When it ended the three of us and Herschel lived in a cabin on Archie Creek and made ties. Together we could make eighty ties a day. Mother continued to run the cookhouse. September rolled around, and Bill and Herschel returned to high school in Mountain View. I continued to work with Dad until January.

More and more news of hard times reached us. The bread lines in cities were getting longer. Men were selling apples on street corners to keep their families from starving. We learned about infant twins, the children of unemployed transients, who had died of malnutrition in Evanston. People were up in arms; the city had to bury the two babies, but food to keep them alive had been unavailable.

I enrolled at the University of Wyoming in January. My funds were limited, of course, and part-time work was scarce. I lived in a basement room in a home in Laramie. In return for taking care of the furnace, getting the landlady's kindling and coal, and shoveling snow off the walk, I got a bargain on the room. I took my meals at the University Commons for the first month and then decided it was too expensive to eat there. I ate scant meals at cheap restaurants the second month.

For the rest of the year I batched with three other students who had an apartment in the basement in which I lived.

There was an abundance of water in Smith's Fork in the spring of 1932, and the drive was over by the time I returned home in June. During part of the summer of that year Dad, Bill, and I lived in a cabin at Gilbert Meadows. The three of us managed to make eighty ties a day which provided us with a wage which was not bad in view of existing circumstances.

One day one of the company bosses stopped to see us and had his noon meal with us. He said to Dad, "You're making a lot of ties, George."

Dad replied, "Yes, we're doing pretty well."

The boss said, "But George, don't make too many."

Every trace of a smile faded from Dad's face. He responded, "Well, God dang it, I ain't goin' to starve."

"Oh, no," said the boss. "I wouldn't want you to do that. But George, don't make too many."

Dad did not say he would or would not make too many, but he repeated, "Well, God dang it, I ain't goin' to starve."

We continued to turn out eighty ties a day until the word came in early July to cut no more as the company would not pay for them until further notice. Herschel had been helping Mother in the cookhouse, but the latter also closed. We parked the ties, built the road, and left Gilbert Meadows.

Dad and Bill found work in a hayfield back in the Piney country. I managed to get a job on the Green River drive. The company had been producing ties on Horse Creek, and ties from the spring drive were at the boom near Daniel. The drive to the town of Green River was launched in July.

One of the men drowned on the drive. The tie hacks were not an unusually sentimental bunch, but several expressed grave concern regarding the sadness and hardship that the man's wife and children would have to endure. One fellow remarked, "Ya, for his family it vill be tough. Of course, for him, hell! He's better off than ve are." Perhaps that expressed the prevalent atti-

(George Linn, 1923)

End of the drive.
Tie boom located at Green River, Wyoming

tude among many tie hacks. They were not despondent and did not appear to be self-pitying. They did, however, seem to expect little more out of life than more and more hard work until they died. With economic conditions as they were in 1932, perhaps they considered themselves lucky to get a living at all; and they took conditions pretty well in stride.

At the end of the summer, Bill went to stay for a time with Granddad on his ranch in the Hoback Canyon. The following winter Bill joined the Civilian Conservation Corps, referred to often as the Three Cs. After the drive ended I worked on a dude ranch for about ten days and then returned to the university. Dad and Mother went back to Smith's Fork and ran the cookhouse for the few fellows who were left in the camp. There were already enough ties cut for the 1933 drive.

Back in Laramie for the fall quarter of 1932, I lived again in a basement apartment with two other fellows. We managed to eat most of the time. At one point our funds were exhausted. All we had in the apartment to eat was a bag of corn meal, so we lived on corn meal mush for about a week. At the end of the quarter, I was broke and thought of going home. Dad wrote and urged me to hold on if I could. There was nothing, he wrote, for me to do on Smith's Fork.

I managed to get some work on the campus during Christmas vacation. I then landed a job working in the home of some people whose business kept them away from home during the day. That was a godsend as I got room, board, and laundry for most of the remainder of the year. I was able to earn only twelve units of credit each quarter at the university, but I had food and shelter and I stayed on. I remained on that job until the last month of the school year when my employers moved to Cheyenne. I lived alone for the remaining month, hopped a freight and went home.

The trip by what was then sometimes called "side-door Pullman" was an experience in itself. In Laramie I climbed to the top of a freight car headed west. Two young fellows who were en route to Oakland joined me. When we reached Rawlins they left me to spend the night at the Salvation Army. Soon I was joined by an older man who was headed for Idaho, hoping to get work in the hay country. We rode on top of the car during the night until we reached Rock Springs. Cinders blew into my face and hair with each puff of the locomotive. The train stopped in Rock Springs, and we found an empty car occupied by about ten other men.

The train started to move again and very soon we were in Green River, a division point on the Union Pacific. Most of the fellows left the train, hoping to find a flop for the night. Two old hoboes and I remained in the car. After about fifteen minutes someone flashed a light through the door. It was the railroad bull, and the car-whacker was right behind him.

"All right," the bull yelled, "what are you doin' in there?"

"Waiting for the train to move," one of the fellows answered.

"Well, you better get off," said the bull. As he shooed us in the direction of the gate to the railroad yard, he said, "Hurry up! I ain't got time to fool around all night."

The car-whacker closed the door of the empty car and sealed it. We left the yards, and I wandered toward the highway thinking I might hitchhike the rest of the way home. It was about one o'clock in the morning, there was little traffic on the road, and it was cool. I walked back to the restaurant near the railroad yard to get a cup of coffee. I saw the two old fellows who had also been kicked off the train. They were standing on the street corner shaking out a tarp in which some of their belongings had been rolled. I asked them if they were going to remain in Green River for awhile.

"No, we ain't," one of them replied. "We're goin' to take that train out of here."

I went into the restaurant and ordered a cup of coffee, and the two fellows came in and also ordered coffee. One moved over next to me at the counter and said, "Listen, kid. We're goin' to board that train when it slows down at the crossing. String along with us, and we'll show you how to git back on."

We went out and waited for the train to be made up. Before long there were twenty-five or thirty fellows waiting and peering into the yard to see whether the train was about ready to move. Some of those old hoboes seemed to know about every junction and division point on the railroad from coast to coast. They also knew at which stations they dared go into the lavatory and wash up and shave. Most of them were traveling in the hope of landing some seasonal work during the summer. Job possibilities and locations where they could best survive were the main topics of conversation. Occasionally there was a little philosophizing.

One old fellow said to me, "You know, I sort of believe that a man gits just what he expects. Now I have earned as much as a hundred and a half a month which ain't bad for a working man. And when I ask for a job and figure I'm goin' to git it, and expect to git it, I git it! If I'm not so sure and don't expect to git it, then I don't git it. Now, sometimes I stop at a house and ask fer a meal. If I expect 'em to give it to me, I git it. If I don't, something seems to tell the consciousness of those people that I don't really expect it, and I don't git it."

As passenger trains sped by we saw car after car loaded with teenagers in the khaki apparel of the Three Cs. Some of the fellows called them the woodtick army. Most of them probably approved of the Three Cs as they knew what it was to be broke and unemployed. One old fellow who probably could not afford to buy a meal, however, was heard to mutter "It's us that's a-payin' for all this."

Occasionally we glimpsed the railroad bull walking along the track checking each car. It was broad daylight by the time the train finally pulled out. As it slowed down at the crossing, I climbed to the top of a carload of hogs.

Fortunately, all freights stopped at Carter to take on water. I arrived there a little before noon and started walking toward Fort Bridger seven miles away. Before long the mail truck overtook me, and I rode to Fort

Bridger and on to Mountain View. I rode another mail truck then to Robertson, arriving there in mid--afternoon. At the general store operated by T. R. Jenkins, I washed some of the dirt and coal dust from my face and tried to comb the cinders from my hair. I had not slept for about thirty hours, and the Jenkins family allowed me to lie down and sleep until evening. I got a ride to the drive camp nearby that night.

I worked on the drive all the way to Granger. As there had not been much activity in the tie camp during the past year, the fellows were all pretty soft; and the boss, Nils Olstrum, did not push any of us too hard. Ernest Nelson proved to be a good cook; and it was a pleasure to eat three square meals a day, including hot cakes or French toast for breakfast. When the drive ended in Granger, I went to work at the boom where ties were being loaded. That lasted until about the middle of July. I worked then with Dad in the timber for the rest of the summer. Bill was still in the Three Cs.

The year of 1933-34 at the university was more comfortable and more successful for me academically. I had a campus job and lived with another fellow during the fall quarter. My wages were thirty cents an hour; but one could buy a pound of hamburger, a quart of milk, and a loaf of bread for twenty-five cents. If we were nearly broke, we ate beans; when we ran out of beans, we ate potatoes. We managed to eat and with occasional help from home to pay our fees and buy our textbooks. During the winter quarter I roomed and boarded at an inexpensive boardinghouse and then landed a board job for the spring quarter. At the end of the year I fully expected to go on the Smith's Fork drive, but the job did not materialize. The worst of a timber company's fears were realized when the snowfall on Smith's Fork was so light that there was not sufficient water to move the ties in the spring.

Times were still hard, and there were plenty of truckers in the Bridger Valley who wanted jobs. Many were hired to haul ties out of the forest to the railroad. In addition to the sound of saws and the rhythmic blows of tie hacks' axes, one would hear the roar of the motors of trucks throughout the forest as they made their way over the crude roads. Occasionally, too, one would hear the explosion of a tire pierced by one of the many tree stumps. It was not an easy operation, but the ties did get hauled.

This may have been the beginning of the end of the tie drives. Trucking ties did not depend on the water level, and weather conditions were of less importance. Drives angered ranchers who needed water for irrigation, and their headgates were frequently damaged by floating ties. As road conditions and the quality of trucks improved, more and more ties were trucked in various parts of the state. The last Wyoming tie drive was the Wind River drive in 1946.

I worked in the timber during the entire summer of 1934. Conditions were better. The company had a new contract with the railroad. Taking advantage of Section 7a of the National Industrial Recovery Act, timber workers had organized. The workday was eight hours instead of ten. Fewer ties could be made in a day, but wages were better.

The company opened a camp on Steele Creek a short distance above the Smith's Fork commissary. Mother ran the cookhouse, and Herschel worked as the flunky. Dad and I made ties. Bill returned from the Three Cs at the end of the summer. Having been out of school for two years, he had perhaps received the short end of the stick; but he valued his experiences in the Three Cs which had enabled him to travel and work in different parts of the country.

Mother, Bill, Jimmy, and I moved to Laramie in the fall of 1934. Later in the year the folks bought a house there, and Mother started providing meals and lodging for university students. Bill and I enrolled in the university, and we each had campus jobs. Herschel, who had just graduated from high school, worked with Dad on Smith's Fork. At the end of the school year we all went back to the western part of the state. I worked at the boom in Granger for awhile and then joined the family at Gilbert Meadows on Smith's Fork. Bill, Herschel, Dad, and I spent the rest of the summer making ties there.

At the end of the summer we returned to Laramie. Dad obtained work with another timber company nearby, enabling him to be with his family more often. He continued to make ties. From 1937 until 1940 he worked as a foreman on Forest Service jobs. With World War II underway, much of the Forest Service work was curtailed in 1941; and Dad found employment as a foreman at an ordnance depot in Oregon. Thus he finally ceased to work in the timber after twenty-one years.

After leaving Gilbert Meadows on Smith's Fork in the fall of 1935, I never again saw a Standard Timber Company tie camp. I did visit a few tie camps in eastern Wyoming in 1937. They presented quite a different picture from those I had known earlier. No broadax was in sight. There was no sound of sleigh bells on teams drawing sleds through the forest. Several bunkhouses were each occupied by about four men, not by eight to twelve as had been customary at one time. There was a bathhouse and a building in which men could hand launder their clothing. While a few old tie hacks continued to hew ties in certain out-of-the-way sections of timber, nearly all of the ties were made at sawmills.

Tractors moved portable sawmills from one setting to another. Bulldozers cleared roads through the timber. Men worked all year falling trees and cutting them into eight-foot logs. Sawmills were operated throughout the summer and until the snow became too deep for them to function. Horses were still used, not for hauling ties, but for skidding logs to the mills. The sawmill crew rolled the logs into place. One man operated the saw, and another turned the log after one side had been sawed.

The heaviest work was done by the two off-bearers. They carried the slabs sawed from the tie and tossed them on a scrap pile. After the tie was completely sawed, an off-bearer shouldered the tie and carried it to the top of a pile. There a man peeled off the remaining bark. The ties were then trucked to the railroad. Sawing ties was more efficient than hewing them, and a log too big to be hewn could be sawed into not just one but two or three ties.

Log cutters worked at a piece rate. Working eight hours a day for six days a week sawmill crew members were paid an hourly wage. With higher wages, more men owned cars and could afford an occasional trip to town. As roads to the highway were kept clear for trucking ties, men could drive to Laramie for a Saturday night and Sunday. Several timber companies operated in the area; and there were tie camps at Horatio Rock, King's Canyon, and near Fox Park. Men and families from various camps attended Saturday night dances at Wood's Landing or at the Vasa Lodge in Laramie.

Social Security had come into existence, and each man had a small percentage deducted from his paycheck. There was much less talk of depression than in 1931, but there were still large numbers of unemployed people. A timber worker ran greater risk of finding himself out of a job than he had a decade earlier. With an increasing number of labor-saving devices at work, there was always a chance that a man's services might not be needed on a sawmill crew; and the company might have enough log cutters. Generally, a good lumberjack could find work; but, if not, he now had unemployment insurance.

There were more families in the newer tie camps. Like the bunkhouses, their log houses contained only meager furnishings and crude uncomfortable furniture. On a timber worker's wages, he could not buy expensive furniture and have it moved to his house only to have it moved again in nine months or so when the family had to relocate in a different stretch of timber. Men continued to bring their own bedrolls consisting of blankets and quilts into the bunkhouses. Living conditions were still far below the post-World War II standards demanded by unions and health departments. Even so, they were a tremendous improvement over those of the early 1920s.

One still heard the Swedish language spoken frequently. A typical camp had an Ole, a Pete, a Carl, a Gus or two, and perhaps an Eric. A slight flavor of the older tie camp remained. With slight variations the same kinds of foods were served in the cookhouse. More fresh vegetables and fruits were available, however. Swedish folk songs were heard occasionally, and Swede and non--Swede alike danced the hambo and schottische.

In general, timber workers and their families seemed pretty well-satisfied in 1937. While jobs in the timber were still far from easy and comforts and luxuries were limited, the tie hacks were far better off than they had been in 1920.

CHAPTER FOURTEEN

Why Did They Do It?

Why would a man live in an isolated camp under such primitive conditions with nothing to look forward to year after year except more of the same backbreaking labor for ten hours a day six days a week? Vital necessities were available only at prices set by the company. There was less luxury available than to workers in cities and small towns. Why would a man continue to hack ties? Perhaps it was to earn a living.

The typical tie hack was probably not qualified for a better-paying job. Work as a general laborer in an urban area was not as difficult, but the pay was less and a man could not count on steady employment. A mine laborer's job was less confining, but it was more hazardous and jobs in mines were not always available. Year-round ranch jobs were limited in number, paid less, and did not provide workmen's compensation in the event of injury. During most of the 1920s the tie hack had a measure of security. He knew he had a job as long as he wanted it. After the 1929 crash employment was scarce, of course, and tie hacks from varying backgrounds and from different parts of the country clung to their jobs for dear life.

There were a few salaried jobs in the timber, but most of the men did not seem to covet them. A drive boss might endeavor to put a drive through in record time in the hope of getting to boss another drive the following year. There were also available the jobs of camp manager, timber boss, and a few other salaried positions. On the whole, there seemed to be little competition for such jobs. Most of the men seemed resigned to their status.

This does not mean that the tie hacks were wholly satisfied. In view of their living and working conditions, they had plenty about which to gripe. They certainly did not always get what they wanted, but a boss never kept them in line by physical force or by throwing his weight around. While a tie hack seldom complained about the price of board at the cookhouse, he demanded his money's worth both in quality and quantity. There was occasional grumbling over commissary prices, but it seemed assumed that nothing could be done about them. Occasionally resentment was expressed at having to cut wood for the cookhouse. This requirement was dispensed with after the move to Smith's Fork in the late 1920s. The company started having the wood sawed with a power saw and chopped by the camp flunky.

From 1919 to the early 1930s if the tie hacks had complaints they kept them as well as their ambitions pretty subdued. At least on the surface they seemed content to work steadily for months and then to draw a check, lay in a supply of liquor, and go on a spree.

When timber workers organized in 1934, however, many grievances surfaced. First and foremost, the men wanted higher wages. In addition to a monthly statement from the company, they demanded a copy of the bill for each purchase showing the prices of items charged at the commissary and a slip showing the number of ties credited to their accounts after each inspection. They also wanted bathhouses and places to do their laundry. When company officials protested that complying with such demands would force the company out of business and there would be no work for anybody, certain of the Swedes said, "Oh, vell, then, ve yust starve to death." The tie hacks did not get everything they wanted, but wages increased and conditions improved substantially after a union was organized.

Some of the old-stock Americans expressed hope to see the day when they could live comfortably without making ties. Perhaps typical of those was a fellow I heard talking with Dad in about 1930. The man had lived most of his life in Oklahoma and Arkansas. He said, "What I want to do is git me a little place with a house and barn, with some pasture, and git me a few old cows, some chickens, three or four hogs, raise me a good garden, and let the ties go to hell!"

Dad's and Mother's primary goal was to get their sons educated and give them opportunities and advantages they never had. At times Dad thought of trying to acquire a ranch in western Wyoming, but he knew the risk involved and the number of years before he could realize much income from such a venture. He possessed the skill for cutting ties, had steady work, and stayed with it.

On just one occasion I heard Dad state what he hoped might lie ahead for him after his sons were grown and educated. He spoke of a day when Mother and he might have a small farm, perhaps in Arizona, which he could operate without hired help, wear overalls and a straw hat, and have a comfortable living. His dream did eventually come true, except that it was in Oregon and not in Arizona.

CHAPTER FIFTEEN

What Happened to Them

Soon after Dad left the Forest Service, Mother and he moved to Oregon where Dad bought an eleven acre place at Hermiston. They kept a few milk cows and raised pigs and chickens. Dad retired about 1959. Mother and he did a little traveling. They became avid rock hounds and gave much attention to their rose garden. Dad died in 1972. Mother, now ninety-two years of age, has lived in California since 1973.

After a few years at the University of Wyoming, Bill passed a Junior Forester's examination with the United States Forest Service. Because of the international situation, however, the Forest Service was not then hiring many people. Bill enlisted in the army in 1940 and served until 1945. After the war ended he went to work for Western Brass Corporation in Los Angeles, and he remained with that company until he retired recently.

Herschel received a degree in accounting from the university. He worked in a payroll office in Hood River, Oregon, until his death in an automobile accident in 1953.

Jim had the musical talent in the family. He took violin lessons for a number of years in Laramie from two university violinists. He also studied with Boris Sirpo, a noted Finnish musician living in Portland. He lost his life in an automobile accident in 1945.

I received my bachelor's and master's degrees from the University of Wyoming and my doctorate from the University of Southern California. I taught speech in various high schools, colleges, and universities, and was also involved in school administration. My work took me to Wyoming, Oregon, Arizona, Alaska, and Washington before I settled in California. From 1955 until 1978 I worked in speech correction and special education programs. In the latter year I retired from my position as Director of Special Education for Ventura County.

It was my good fortune to sail on the **Kungsholm,** a Swedish-American Line ship from New York to Sweden in 1958. We pulled into the harbor at Gothenberg shortly after two o'clock in the morning. Probably every passenger on that ship got out of bed and went up on deck. In addition to those of us who were getting our first glimpse of Europe, there were people who had sailed from this dock thirty years or so earlier who were returning to their native land for the first time in many years. Some were hugging each other and almost weeping. A number were excitedly pointing out landmarks, including the elevated statue of the sailor's wife looking out over the bay.

I thought of so many people who had embarked from that very port for the new world and a new way of life: Edins, Johnsons, Molines, Stranquists, Larsons, Ericksons, and ever so many others. I recalled Andrew Ostlund and Victor Latine, two of the Bloom camp employees, who had returned to the old country in 1921. Andrew married his sweetheart who had remained in Sweden. Hoping to return to his wife and family in Sweden with a stake, it had taken Victor eight years to accumulate the latter. Frank Edin, my first drive boss, moved his family to the old country in 1934.

With a few exceptions the Swedish children we knew had no opportunity to visit their parents' homeland. Linnea and Alfred Moline spent four years in Sweden after their mother died. When they returned to the United States they continued to observe Swedish practices. In fact, Alfred would lift his hat when greeting not only a lady but also a man or another boy. Nils Edin visited Sweden with his parents in the summer of 1928. He reported that he had no firecrackers on the Fourth of July. After being in Sweden for awhile he wanted pie, but their relatives with whom they stayed had no pie tins. At last Mrs. Edin made him a pie, using a milk pan.

Most of the Scandinavian immigrants settled down to life in America and eventually took out their naturalization papers. I seldom heard a Swede express any hope regarding a future goal or aspirations for the future. A number had savings accounts in banks or with the building and loan companies which flourished during that era. It appeared, though, that a Swede's reason for saving his money was to have funds for a rainy day. He held no hope that such savings would enable him to find another means of getting a living. I have heard of very few of these lumberjacks who eventually acquired farms or ranches or went into business. The great majority seemed to expect to hack ties indefinitely. Among those who were still working at the time of their deaths were Gus Satline, Nils Nasland, and John Pearson.

They escaped the fate of Carl whose cabin was next to ours during our year in the Moline camp. After Dad built his cabin, old Carl lived with him while he built his own cabin. Carl, a widower, made ties. He was quite a good cook. During the third year of the Moline camp, the cook left suddenly and Carl ran the cookhouse until spring. He liked to garden. In the late spring he fenced and spaded a plot of ground near the creek and grew vegetables. In 1925 Carl was no longer physically able to hack ties. When I last heard of him he was living on a ranch where I suppose he did chores for his board. He drew twelve dollars a month from the county, and he was then raising a garden on which he hoped to net one hundred dollars.

Others among the tie hacks continued to work for many years, prospered to a greater or lesser extent, and their families entered the mainstream of American life. John Moline, who we first knew in the Moline camp in 1919, hauled ties for the company for many years. He spent his last years in Oregon. His older son Andrew lives in Saratoga, Wyoming, and his younger son Alfred recently died in Oregon. Linnea, his daughter, lives in California. Chris Conchef, who was at the Bloom camp in 1920-21, left the timber about a year later and went to work in the mines near Rock Springs.

Bror Edin, the camp manager all those years, died during the mid-thirties. His widow and son later located in Oregon where Mrs. Edin lived until her death in 1976. Their son Nils received his degree in forestry from Oregon State College and resides in Oregon. Gus Gill worked intermittently for the company from 1919 until 1935. After the death of Bror Edin, Gill served for a time as camp manager. I have heard that he and his wife are living in retirement in Oregon. My second drive boss, Elmer Johnson, was reported to be living in Idaho where he operated a mink farm. Nils Olstrum, my third drive boss, was still working for the company in 1936.

Andrew Bloom, a teamster, worked for the company until 1926. In 1937 I heard that he was working at the Wind River tie camp for a different company in northwestern Wyoming. Walfred Peterson continued to work for the company for many years and died during the 1940s. Many of the men left the Smith's Fork tie camp during the big exodus of 1930. I have no idea where most of them went, how they got a living, or what happened to them.

"W" moved into a house near Robertson in the late 1930s. Over a period of time he became increasingly interested in automobiles. A Forest Service man observed that if a good automobile salesman got hold of "W," he would sell him a car. After "W" retired he did buy a car. One of the questions he reportedly asked the salesman was "How many horsepower is best for getting out of a mudhole?" I do not know what answer he got, but he bought the car. "W" lived until 1946.

One who experienced tie camp life enjoys reminiscing about it. Life was simpler, but was it easier? I have heard young people today assert that a laborer of the earlier days was really better off. While the hourly wage was less, they contend that without the Social Security tax, income tax, union dues, and with lower prices a man at that time really came out ahead of today's worker. People who make such statements did not live and work in a tie camp. Too many of the tie hacks I knew endured deplorable conditions, worked harder than I would want anyone to have to work, and got very little in return.

During the most prosperous times even the best tie hacks could hardly expect to save enough for a comfortable retirement. Many continued to make ties when they were quite aged and could barely subsist on what they earned. A man who became too feeble to work at all was obliged to live in poverty.

Yet to me there was a time when life in the timber seemed good. I still have fond memories of going home from school in the late afternoon, sawing and splitting wood, studying or reading by the old woodburning heater after a hot supper, and then crawling into a warm bed and hearing the wind whistling through the trees. When we were in high school it was always good for the school year to end so we could return to the tie camp. It was equally good, though, to have the summer end and get back to town. As we grew older we became more and more aware of the advantages of life with electricity, piped-in hot and cold running water, comfortable furniture, and more leisure.

To spend a few weeks in the timber today would be comparable to a prolonged hunting trip or perhaps an exploring expedition. To continue to live there indefinitely would become discouraging and stale. To do the type of arduous rugged labor required of the tie hacks of yesteryear with so little return and no hope for a better future would be discouraging. Few today would tolerate it.

Good, bad, or indifferent, the tie camps in which I grew up are now a part of history. No longer on Cottonwood Creek or Smith's Fork does one hear the crash of a tree as it falls to the earth, the rhythmic sound of a tie hack's ax, or the sleigh bells as teamsters drive through the forest. Logs are now cut in the forests with power saws. They are transported to modern sawmills and sawed into ties which are hauled by truck or train to where they are needed.

Gone also are the chinked and daubed log cabins and bunkhouses with their rough lumber doors and wooden latches, smoke pouring from the stovepipes, and distinctive smell of wood smoke mingled with that of tobacco smoke and snoose.

Perhaps the greatest reason for my fond memories of that era is my admiration for the tie hacks who were broadax artists. They were not all alike. Most of them were thrifty but generous, but a few were moochers and deadbeats. A number of them were as clean as they could be while living in a crowded bunkhouse, and a few were positively filthy. Most were steady, dependable, and hardworking. Some were chronic drunkards and loafers. The great majority were friendly and extremely cooperative, but a few were hostile and belligerent. Most were honest, but some were not. Of one man, old "W" said, "I vouldn't trust him vith a dead cat."

All in all, no one can deny that the tie hacks of the 1920s and 1930s were a robust hardworking lot. I admire and respect them. They are among the unsung heroes upon whom America's industrialization and progress depended.

Index

About the Author

Born in western Wyoming, George B. Linn grew up in tie camps and obtained his early education by home study and in one-room schools. While in high school and university, he worked in the timber during the summers. He received his bachelor's and master's degrees from the University of Wyoming and his doctorate from the University of Southern California. His professional life was devoted to teaching speech courses, speech correction, and school administration. Since retiring from his position as Director of Special Education for Ventura County in 1978, he does research in language development and history of the American West. He has written articles and reviews for professional journals.